Managing the Return of the Wild

This book explores attitudes and strategies towards the return of the wild in times of ecological crisis, focusing on wolves in Europe.

The contributions from a variety of disciplines discuss human encounters with wolves, engaging with traditional narratives and contemporary conflicts. Covering a range of geographical areas, the case studies featured demonstrate the tremendous impact of the return of the wolf in European societies. Wolves are a keystone species that exemplify humanity's relation to what is called nature and their return generates powerful debates about what 'nature' actually is and how much it is needed or should be permitted to exist. The book considers the return of the wild as a catalyst for fundamental socio-biological changes of the world within human societies, and the various responses of humans to wolves demonstrate both our potential and limitations when it comes to multispecies communities and negotiating societal change.

Managing the Return of the Wild will be relevant to a broad audience interested in discussions of social and ecological conflict today, including scholars from multispecies studies and diverse disciplines such as biology, forestry management and folklore studies.

Michaela Fenske is a professor of European Ethnology at the University of Würzburg, Germany.

Bernhard Tschofen is a professor at the Department of Social Anthropology and Cultural Studies (ISEK) at the University of Zurich, Switzerland.

Managing the Return of the Wild

Human Encounters with Wolves in Europe

Edited by
Michaela Fenske and Bernhard Tschofen

Routledge
Taylor & Francis Group

LONDON AND NEW YORK

First published 2020
by Routledge
2 Park Square, Milton Park, Abingdon, Oxon OX14 4RN

and by Routledge
605 Third Avenue, New York, NY 10017

First issued in paperback 2022

Routledge is an imprint of the Taylor & Francis Group, an informa business

Publisher's Note
The publisher has gone to great lengths to ensure the quality of this reprint but points out that some imperfections in the original copies may be apparent.

British Library Cataloguing-in-Publication Data
A catalogue record for this book is available from the British Library

Library of Congress Cataloging-in-Publication Data
A catalog record has been requested for this book

ISBN: 978-0-367-49873-3 (pbk)
ISBN: 978-0-815-35341-6 (hbk)
ISBN: 978-1-351-12778-3 (ebk)

DOI: 10.4324/9781351127783

Typeset in Times New Roman
by Taylor & Francis Books

Contents

Figures

Contributors

Irina Arnold is a doctoral fellow in the research project "The Return of Wolves: Cultural-anthropological studies dealing with the process of wolf management in the Federal Republic of Germany", funded by the German Research Foundation (DFG) at the Institute of European Ethnology at the University of Würzburg. Based on a case study in the federal state of Lower Saxony, she is investigating the interactions and relationships between wolves, humans and other non-human animals.

Filipa Costa is a doctoral fellow in the Science and Technology Foundation (FCT) PhD Programme in Anthropology: Politics and Displays of Culture and Museology (ISCTE/NOVA/CRIA). Her current work focuses on a Portuguese National Park, its visitors and non-human beings. There, she is exploring the processes of natural heritage production and the related concepts of nature and culture.

Rony Emmenegger is a political geographer and postdoctoral researcher at the Sustainable Development Group at the University of Basel. His current interests lie with the governance of risks in Switzerland and the ways an advancing 'risk society' engages with unruly natures and materials. His work provides insights into the politics of human–environmental relationships and the geographical interdependencies of the relationships between state, society and nature.

Meret Fehlmann is the head librarian of the library of popular cultures and lecturer in Popular Culture Studies at the Department of Social Anthropology and Cultural Studies (ISEK) at the University of Zurich. She is interested in popular literature and media with a particular focus on gender aspects and human–animal studies. Since 2011, she has been co-editor of the online journal *kids+media*.

Michaela Fenske is a professor of European Ethnology at the University of Würzburg. Her current research interests include the Anthropology beyond the human/Multispecies ethnography. She supervises the research project "The Return of Wolves: Cultural-anthropological studies dealing with the process of wolf management in the Federal Republic of Germany", funded by the German Research Foundation (DFG).

Elisa Frank is a doctoral fellow at the Department of Social Anthropology and Cultural Studies (ISEK) at the University of Zurich, where she works in the research project "Wolves: Knowledge and Practice. Ethnographies on the Return of Wolves in Switzerland", funded by the Swiss National Science Foundation. Interested in human–environment relationships, her research focuses on representations and discourses on and about wolves in contexts exceeding the official, institutionalized wolf management and the immediate core conflict of agriculture vs. nature conservation.

Thorsten Gieser is a lecturer in Anthropology at the University of Koblenz-Landau. Interested in Environmental Anthropology and Cultural Phenomenology, he has conducted research on hunting, focusing on human–animal relationships, landscapes and the senses and sensibilities associated with them. He is currently investigating the return of the wolf to Germany within the framework of the research project "The Return of the Wolf to Germany: mapping extraordinary affective encounters" funded by the Volkswagen Foundation.

Marlis Heyer is a doctoral fellow in the research project "The Return of Wolves: Cultural anthropological studies dealing with the process of wolf-management in the Federal Republic of Germany" funded by the German Research Foundation (DFG) at the Institute of European Ethnology at the University of Würzburg. Based on a case study in the region of Lusatia, she is conducting research on narratives and narrations of the return of the wolves, wolf management and the coexistence of humans, wolves and other non-human beings, considering historical as well as current discourses.

Nikolaus Heinzer is a doctoral fellow in the research project "Wolves: Knowledge and Practice. Ethnographies on the Return of Wolves in Switzerland", funded by the Swiss National Science Foundation at the Department of Social Anthropology and Cultural Studies (ISEK) at the University of Zurich. His interest lies with people living near wolf territories in (mountain) areas of Switzerland in order to examine how the everyday physical presence of wolves is dealt with.

Susanne Hose is the deputy director of the Sorbian Institute in Bautzen, Germany. She is interested in historico-cultural research with a focus on narrative cultures. She has conducted research on disaster narratives in Lusatia, ego-documents as cultural heritage, and tales of the Upper Lusatian heath and pond landscape.

Ilona Imoberdorf is a human geographer and former research fellow at the Department of Social Anthropology and Cultural Studies (ISEK) at the University of Zurich, where she studied the communication of authorities in the presence of "not shy" wolves in co-operation with Carnivore

Ecology and Wildlife Management (KORA). Inspired by human-animal encounters in rural and urban areas, she is currently working as a coach and environmental educator.

Helena Ruotsala is a professor of European Ethnology at the University of Turku. Her research deals with reindeer herding in Finnish Lapland, the different actors involved and local landscapes. Her work provides an insight into modern reindeer management, the emerging conflicts between the different groups involved and the importance of knowledge transfer.

Laura Siragusa is a linguistic anthropologist at the University of Helsinki, where she currently manages the research project "A tangible heritage: Vepsian language and non-human agencies to co-construct a northern environment", funded by the Kone Foundation. Her research interests are related to the revival of heritage languages, language ecology, verbal art and its relationships to non-human animals and other beings, domestication and health.

Anke Tonnaer is an assistant professor of Anthropology and Development Studies at Radboud University, Nijmegen. She has conducted research on ambiguous tourist imaginaries of wilderness in the Netherlands and Australia.

Bernhard Tschofen is a professor at the Department of Social Anthropology and Cultural Studies (ISEK) at the University of Zurich, specialising in the ethnography of space. His current research interests include Alpine tourism, culinary heritage as cultural property and the anthropology of wildlife. He supervises the research project "Wolves: Knowledge and Practice. Ethnographies on the Return of Wolves in Switzerland", funded by the Swiss National Science Foundation.

Acknowledgements

The starting point of the present volume was the panel "The Return of the Wild: Fears, Hopes, Strategies" that took place at the conference of the European Association for Social Anthropologists (EASA) in Milan in 2016. The editors thank the conference organisers for having accepted this panel as part of the programme. We appreciate the collaboration of a number of colleagues who were not able to participate in the conference but who were willing to contribute to the anthology. We gratefully acknowledge the financial assistance of the organisations funding two of our research projects dealing with the return of the wolves to Central Europe. The Swiss National Fund (SNF) supports the research project "Wolves: Knowledge and Practice. Ethnographies on the Return of the Wolves in Switzerland", supervised by Bernhard Tschofen at the Department of Social Anthropology and Cultural Studies (ISEK) at the University of Zurich. The German Research Foundation (DFG) funds the research project "The Return of the Wolves: Cultural-anthropological studies dealing with the process of wolf management in the Federal Republic of Germany", supervised by Michaela Fenske at the Department of European Ethnology at the University of Würzburg. With their financial support, two teams of junior researchers, Elisa Frank and Nikolaus Heintzer in Switzerland and Irina Arnold and Marlis Heyer with student assistant Laura Duchet in Germany, are able to pursue their studies concerning the effects relating to the return of the wolves. The Sorbian Institute in Bautzen, represented by Susanne Hose, is a partner to the German project, since the region of Lusatia was one of the first regions in Central Europe where new experiences with wolves were gained in the 1990s. Both research teams work closely together, and the present volume is a result of this collaboration. Last but not least, the editors acknowledge the collaboration of translator and proofreader Philip Saunders, Berlin. Over the past few years, several English language publications in our institutes profited considerably from Philip's commitment and expertise.

1 Human encounters with wolves

An introduction

Michaela Fenske and Bernhard Tschofen

Of bees and wolves: the return of the wild in times of ecological crisis

Swiss filmmaker Markus Imhoof narrates the mass dying of the European honeybee at the beginning of the twenty-first century in his prize-winning documentary, *More than Honey* (2012). The widely popular documentary deals with the severe health problems honeybees experience in different parts of the world. Aiming to explain the mass dying of honeybees, scientifically assessed as colony collapse disorder, Imhoof identifies one of the main reasons as human influence, particularly the ruthless capitalist exploitation of natural resources. Meanwhile, in China, bees and other insects are so rare that humans have taken to pollinating the plants themselves. The filmmaker uses this example to evoke the dramatic image of a world without honeybees. Imhoof's film exemplifies today's human anxiety in the face of nature falling silent that has been identified by German sociologist Hartmut Rosa (2016, 463f.). Rosa reflects the phenomenon as one aspect of a dramatic loss of human entanglement within the world (*Verlust an Weltbeziehungen*). At the same time, at the beginning of the twenty-first century, a growing number of people share an increased awareness of being part of an entangled world of living beings, humans and 'other-than-humans', in which the latter part is becoming increasingly lost (Wright and Simpson 2014).

Imhoof's film ends on an optimistic note, as he introduces attempts to breed resilient honeybees. A group of Australian scientists succeeded in breeding a hybrid of domesticated honeybees and wild bees. The hybrid bees are set free on an island far away from human civilisation so as to ensure that they will not harm humans. In their own words, the scientists intend to ascertain that the hybrid bees will not pose a similar danger for humans as wolves will do, because humans "cannot live together with wolves" (Imhoof 2012). Through the backdoor, Imhoof's popular narration of ecological crisis evokes familiar dichotomies. On the one hand, there are the honeybees that have been domesticated and praised by humans for their virtues since antiquity. On the other hand, there are the wolves, the prototypical antagonists of human civilisation that humans have feared and persecuted for centuries. These animals represent the stereotypical sympathetic and antipathetic, the so-called domesticated and wild fauna. At a time when perspectives and orders developed in modernity have become increasingly questionable in the face of dramatic environmental developments such as climate change and the extinction of species, Imhoof's film reflects the traditional dichotomy of the domesticated and the wild. The former is usable by humans but subject to severe damage by disease, the latter is fairly resilient but, at the same time, experienced as constituting a severe threat to human civilisation. The essential message of this narration is old-fashioned but still popular: the domesticated and the wild, culture and nature, should be clearly separated. If they are merged, this can be tested safely only on an island far away from any human dwellings so as to protect civilisation.

At the beginning of the twenty-first century, honeybees and wolves are regarded as keystone species, both of them attracting considerable attention in European societies. The one species because of its mass dying that is representative of the "insectageddon" of the twenty-first century (Monbiot 2017). The other because of its revival that, in relation to the relatively small number of individuals, receives enormous attention almost everywhere in Europe. Just as humans discuss the consequences of the ecological crisis of the twenty-first century and their own impact on the planet under the term 'Anthropocene', wolves reappear. Wolves were exterminated in most parts of central Europe, except in some peripheral areas where they have survived.[1] At the same time, the species constitutes an extraordinarily effective actor in myths, fairy tales and other popular narrative genres. And now, for about three decades and increasingly so in recent years, the species is recovering. As the eviction of the wolves was effected by humans, so their return is also partly due to human influence. It is due to wildlife conservation and laws and rules of species protection that increasing numbers of wolves have existed within European landscapes since the end of the twentieth century.

At the beginning of the twenty-first century, the wild and the civilised are no longer separated, and – as the Australian scientists in Imhoof's film point out – most people in European societies are not (or no longer) familiar with living together with wolves and other wild animals. Sheep breeders, hunters and villagers complain of wolf attacks, and although the species is protected by law, politicians feel obliged to ask for restrictions to the protection of the rare animals. Encounters between humans and wolves engender various conflicts that are transported into the public discourse by seemingly endless stories throughout the media.

In addition to the dimension of conflict, other dynamics are clearly evident. The wild is also seen as an important resource for the recovering of weakened domesticated species, such as honeybees, or for the recovery of ecosystems that do not function. Therefore, for some Europeans, the return of the wolves also constitutes a sign of hope that recovery is still possible in the current "catastrophic times" (Stengers 2015). Nevertheless, the problem that needs to be solved is how Europeans in their densely populated and intensely used agricultural areas manage to live together peacefully with the wild new-comers. Citizens, villagers, politicians and scholars are searching for suitable answers concerning the challenges created by the return of the wolves. It has become a topic of great social relevance: a topic of the feuilleton, intense debates in online media and an object of mediating exhibitions in not only classical museums of natural history. Last but not least, the arts are increasingly dealing in a creative way with related social irritations and how the return of the wolves seems to undermine the orders believed to be reliable. To mention just a few examples from recent German-language literature alone: a lone wolf, coming from the East, crosses the border river on a cold night in January, stirring up not only the capital city Berlin and its inhabitants, but also the city's relationship to its surroundings (Schimmelpfennig 2016). Even where a young woman has to guard a factory abandoned due to globalisation and outsourcing of production against the intrusion of wolves, other disturbing things suddenly happen – an unknown man falls out of an airplane and moves the place into the uncertainties of global entanglements (Molinari 2018). And in the crisis-ridden and wolf-infested province of East Germany, a strange trapper (*Fallensteller*) appears who does not catch animals (on the contrary, attracts them) but listens to people with their worries and needs: "We are becoming less, the animals are becoming more" (Stanišić 2016, 171).

The discussions and reactions show to what extent wolves are able to undermine the seemingly fixed borders of European modernity and, thus, question long-standing self-evident concepts, such as hunting or species protection. The ability of wolves to overcome de facto more or less unnoticed great distances, negate 'natural' and material boundaries and adapt to changing environmental conditions, works towards this irritation and, simultaneously, accounts for a substantial part of the emotions – both negative and admiring – in public debates.

In this way, the agency of the predators, which always appear as symbolic actors too, touches inevitably on fundamental (bio-)political questions and the value systems and policies associated with them. Finally and above all, the presence of wolves reveals the contradictions between the main concepts of contemporary societal orientation. The irritation provoked shows how far they have inscribed themselves into the material and immaterial life-world. The desire for security (biosecurity) and the simultaneous commitment to biodiversity become contradictory.

Wolf management, as the management of wildlife as a whole, has traditionally been assigned to the natural sciences. Biologists and wildlife conservationists have been responsible for the protection of wild animals and their control. Today, the return of the wolves and other wild animals, such as the bear or lynx, is also a topic for the humanities and the social sciences. These disciplines emphasise, for example, that bears, wolves and other wild animals are important parts of human cultural heritage (see Nevin, Convery, and Davis 2019). This implies that human civilisation was also formed by wilderness and its creatures and that the latter play a pivotal role in human history and memory. At the same time, scholars in the humanities and social sciences have developed new perspectives that make the very idea of managing the wild itself appear questionable.

Managing the wild? New perspectives

Considering the return of the wild as a catalyst for fundamental socio-biological changes of the world within human societies, questions of wildlife management as part of broader reflections on today's ecological crisis are now an arena of interdisciplinary interest. Consequently, there is a plurality of different voices that also debate essential theoretical concepts: Are modern concepts of wilderness and wildlife protection in the Anthropocene still fruitful in a world in which every single spot has been subject to human influence?

Defenders of the concept of wilderness, such as philosopher Mark Woods, propagate the need for environmental or wilderness ethics that emphasises the "freedom of the other-than-human world" (2017, 242). Others, such as legal scholar and ethnographer Irus Braverman, point more to the plurality of current protective actions and the implied ethical dilemma in times of mass extinction (2015). Seen from a distance, many of the differently voiced positions share a certain awareness of the fact that modern dichotomies are no longer plausible. Neither is there anything such as pure nature and pure culture that can be clearly separated from each other, nor did this dichotomy ever exist. In simple words, the problem is more complex.

If the established separation of culture and nature no longer works, the task of conservationists needs to adapt, as geographer Jamie Lorimer argues (2015). From his point of view, wildlife protection ought to be seen as a democratic process including humans as well as animals and other-than-human beings. Caring for wildlife is increasingly interpreted as a field of experiment in which both knowledge and experience of each member of the multispecies communities concerned has its effects (see also Lorimer and Driessen 2014). Essentially, this also implies the end of any idea of management, since management arises from the illusion that humans might be able to control wild animals, itself an idea of modern times.

A growing number of concepts, especially in the humanities, deal theoretically with the entanglement of animals, plants, humans and other living beings for wildlife studies. Human–wildlife interactions are seen through "a multispecies lens" (Aisher and Damodaran 2016, 6), implying that they are influenced by phenomena researched by multispecies studies (c.f. Kirksey and Helmreich 2010; Multispecies Editing Collective 2017; van Dooren, Kirksey, and Münster 2016). Multispecies Studies or Ethnography (and Human–Animal Studies as part of this broad research field) wants to overcome a traditional understanding of ecology that aims at the conservation of unaffected ecosystems. Instead of deploring the loss of wilderness (if wilderness ever existed), multispecies scholars are interested in understanding current multispecies communities in their mutual dependencies, logics and possibilities. They observe new emerging ecologies, in which so-called invasive species play a role, as do returning species such as wolves and bears (Kirksey 2015). According to multispecies scholars, humans ought to learn to let themselves be influenced, as scholar of environmental studies Sarah Whatmore argues (2019). They should try to adopt the perspectives of other-than-humans and be aware of the need of reciprocal communication and exchange (see also Kirksey 2015). In other words, humans should realise that they are just one actor among others within their living environments. Insights such as these have led to an intense debate, particularly in the humanities, concerning the ethical consequences and demands of the new perspectives (e.g. Altex 2019; Bovenkerk and Keulartz 2016).

The perspectives adopted in this way make the return of the wolves a topic of culture and society. It cannot be understood in its reactions and conflicts without considering the cultural situatedness of such far-reaching concepts as nature and wilderness or tradition and security. The more recent Human–(Non-Human)–Animal Studies, an interdisciplinary field of research in which the study of animals is deliberately not limited to human attributions in artistic and everyday representations or the level of symbolic meanings in various systems of meaning, stands for such an approach (Marvin and McHugh 2019).

In their cultural theory orientation, Human–Animal Studies is based primarily on the principles of gender and body studies and adopts the removal of the separation of nature and culture as laid down in feminist and post-colonial theory (Butler 1993). This international research direction, which is partly interwoven with animal-ethical movements, locates the construction of a natural otherness in the historical-social process and questions the fundamental passivity of the animal, respectively, non-human counterpart, regardless of the power relationships involved. The focus is, therefore, on the question of animal power of action in human–animal relationships, in which borders, orders and, thus, identities are assumed to be open and negotiable. Humans and animals – the plural is important – co-constitute each other in such a perspective (Haraway 2008) and, above all, not according to universal principles but based on situational arrangements. These ideas of a consistently relational anthropology are taken even further in the conception of multispecies ethnography (Kirksey and Helmreich 2010). Following the actor-network theory and its research, especially on knowledge as a hybrid network of social and material components, the focus is not only on humans, non-human animals and other living beings but also on techno-social and discursive elements as powerful entities.

In such a perspective, the current growing expansion of the wolf population is also not simply to be understood as a return of the big predator that was exterminated in most European countries in the eighteenth and nineteenth centuries. What appears to be a natural process and question of biology cannot be separated from the conditions under which wolves find their place in the ecosystem again. For this reason, the image of a simple return is somewhat distorted, since they return to an environment that is completely different from that of the time of their extermination, a space that is now not only industrialised, intensively developed and, in most cases, much more densely populated. But also to a space in which, despite the expansion of the areas claimed by humans, there is now much more forest and much larger numbers of huntable game than at the beginning of the nineteenth century. If animal and human being define themselves only in their relationship to each other, one could even say that the wolf of the twenty-first century is a different animal – in principle, perhaps no less 'historical' or 'social' than humans with their equally environmentally changed ways of life and thinking.

In addition to existing conceptual differences, scholars from different disciplines are, therefore, in favour of new approaches towards (what only some of them still call) wildlife management. Following the aim of learning to live together with other-than-humans, interdisciplinary research underlines the need for a more locally sensitive and place-based care for wildlife that had already been identified by anthropologist John Knight (Knight 2000, 5; see also Aisher and Damodaran 2016) 20 years ago. Scholars argue increasingly for a stronger flexibility within this process (Nevin, Convery, and Davis 2019), pointing to the fact that many (if not most) wildlife conflicts can better be understood as "conflicts between human groups" (Hill 2017, 1). Consequently, anthropologist Catherine Hill argues that "the future for conservation for many wildlife species relies not just on innovative solutions, but also on increased tolerance and social carrying capacity that cannot be achieved by laws, science, money, fences alone" (Hill 2017, 11). Humans will also have to relearn to tolerate their own ambiguities towards wildlife. Moreover, they should not expect simple solutions concerning ways of living together (Fenske and Heyer 2019).

Living together with other-than-humans not only needs new ethics and perspectives but also new practices in everyday interaction. What opportunities do these new ethics, perspectives and practices within the everyday contact zones of human and wild animals offer? One needs, first and foremost, more detailed knowledge about the experiences within local-regional multispecies communities to understand the current situation of human–wolf interaction in European landscapes and to find solutions for a coexistence. This anthology aims to contribute to this process by presenting case studies from different European regions and places.

Encounters with wolves: current research

The present collection of essays results, in part, from the presentations of the panel "Return of the Wild: Fears, Hopes, and Strategies. Ethnographic Encounters in Wildlife Management in Europe", convened by the editors at the 14th biennial conference of the European Association of Social Anthropologists, "Anthropological Legacies and Human Futures". The panel and the publication are inspired by the field of Multispecies Studies and other post-humanist approaches. The disciplinary backdrop of the contributors lies in the anthropologies (including social anthropology and cultural anthropology, such as European Ethnology), folklore and geography, and their international research interests cover regions in countries as diverse as Finland, France, Germany, the Netherlands, Portugal, Russia and Switzerland.[2] The book's contributions study the fears, hopes and strategies relating to wolves in those regions to which the animals are currently returning. Most of the present studies are part of larger research projects focusing on the return of wolves in different European landscapes. All of them are based on empirical research done in archives, libraries and the internet, and they most often result from intense ethnographic fieldwork including collaboration with multispecies communities in different regions all over Europe.

As mentioned above, the title of this book deliberately refers to the traditional concept of wildlife management. However, it expands its understanding and conceives management in a much broader sense as dealing with the return of the wild in politics and everyday life and includes ways of thinking and imagining as forms of action. These are not limited to the direct management of wildlife but also include ways of handling the phenomenon in fields that do not appear to be directly affected by it. This includes, for example, questions of storytelling, mediation in environmental education or museums, and the general reorientation of life, dwelling and working in rural areas.

Accordingly, the perspective of the essays gathered here is an interdisciplinary one, which, by its very nature, is not limited to the conventional approaches of anthropological subjects to environmental issues. Instead, it expands its view of not only the suggestions outlined by multispecies studies but also, for example, general approaches from environmental humanities and science and technology studies. They are united by a basic relational attitude according to which, on the one hand, nature can only be understood in its cultural situatedness and, on the other hand, the connections between politics and practice, knowledge and action can be grasped conceptually. In doing so, the essays take different perspectives, but the following focal points and questions arise which connect the contributions at different points. At least three of these perspectives will be briefly outlined here.

Firstly, the reconstructions of the various historically and ontologically situated experiences and perceptions of wolves should be mentioned. The contributions aim to understand not only human ways of dealing with wolves but the entire more-than-human figuration both as a product and prerequisite of a dynamic relationship. They thus make an important contribution to the de-essentialisation of the wild and, in this case, especially "the wolf" in this widespread but complexity-reducing singular.

Secondly, core questions of the humanities and especially of anthropology come to the fore in the essays on the relationship between knowledge and practice. A number of contributions focus on heterogeneous knowledge and, thus also on conflicts between different orders of knowledge and their spatial figurations, for example, between expert knowledge and local common-sense knowledge. They are able to reconstruct the different access to sources of knowledge and the associated power relationships, and they can, therefore locate the strategies of negotiation and justification of practical action in dealing with wolves.

A third focus combines these two perspectives and puts the relationships between language and narration and the dimensions of knowledge and practice at the centre of its interest. Various essays focus on or address, among other aspects, the associated recourse to cultural memory and explore how talking about wolves constitutes biosocial orders and offers explanations for positions and experiences in the encounter with the new wildlife.

The contributors study how humans experience the effects of globalisation, the developments of their native regions and the possibilities of their futures by reflecting the return of the wolves. In doing so, they talked to the people concerned who are debating the orders of their changing worlds, the status of nature and culture, wild and civilised, and their knowledge about the other-than-human worlds.

In the aftermath of ecocriticism and literary animal studies, popular narrative culture is no longer seen as simply reflecting the world but also as media whose contribution to creating the world should not be underestimated (Borgards and Pethes 2013; Zapf 2016). Popular narrations are of great importance not only for studying current experiences but also for developing future possibilities of human–wildlife encounters. In recent times, for example, there has been an impressive revival of Little Red Riding Hood stories all over Europe, many of them following the tradition of negative stereotyping of wolves formulated by the Brothers Grimm at the beginning of the nineteenth century (Shojaei Kawan 2003). At the same time, urban cultures particularly offer new versions of the traditional story that develop positive human–wolf encounters (Fenske 2018). It is this power of variation of narratives (and related to this, the chance of developing different perspectives even when using the same elements), that is at the centre of the contribution by Swiss folklorist Meret Fehlmann. Following the different versions of the premodern narration of the Beast of Gévaudan until today, Fehlmann demonstrates how the story that has developed the role of the beast has been represented in different ways depending on the related contemporary perspectives concerning wildlife.

The contribution of Marlis Heyer and Susanne Hose deals with Lusatia, an important area of wolf-human interaction in the East of Germany, and the collective memories of the human population. The authors demonstrate how, in the face of wolves, yesterday's fears are updated from collective memory to become decisive arguments in today's politics.

From the perspective of the Netherlands, where a migrating wolf engendered a certain number of narrations in the media, Anke Tonnaer introduces debates on the imaginations and legitimacies of a "new wilderness". Here, the return of the wolves epitomises the ambiguities of the current relationship between humans and their living environments in the densely populated Western world.

Filipa Costa observes this ambiguity towards wild animals in her field in Portugal, where locals and tourists shared their wolf stories with her. Costa's contribution offers insights into both positive and negative feelings towards the return of the wild, the revival of traditional rivalries between shepherds and wolves, and a new valuation of the wild as cultural heritage through tourism.

Elisa Frank's contribution is the first of three case studies from Switzerland. Frank focuses on environmental education. By analysing current settings of environmental education in museums and zoos and through hiking tours, she introduces new understandings of the ways in which contemporary humans both learn about and construct what they define as the environment.

Considerable discordances concerning the meaning and context of wilderness are at the centre of Nikolaus Heinzer's contribution. Based on his exchange with people in the Swiss Alpine regions, Heinzer proves how important the involvement of humans is concerning their perspectives and positions towards sharing their native regions with wolves.

The essay by Ilona Imoberdorf and Rony Emmenegger deals with concepts of space and positions towards wild returners and how these are constructed in different social milieus. Their region is also the Swiss Alps, where they researched intensely the process of constructing the region as a special area for different uses, including human identification.

Irina Arnold follows the traces of the wolf known as MT6, who became popular as Kurti. The wolf was killed a few years ago in northern Germany because of its "problematic behaviour". By using the concept of animal agency in "getting closer", Arnold argues for a more sensitive and human-decentred approach to individuals such as wolves and their effects within multispecies societies.

Hunters play an important role in discussions concerning the return of the wolves. The different attitudes and conflicts in this milieu are discussed by Thorsten Gieser. He deals with conflicts between so-called new ways of ecological wildlife management and the *Hege* practiced by traditional hunters in Germany. *Hege* means a care for wildlife that relies on specific conceptions of both the human–animal relationship and the wild.

The final two essays deal with experiences of humans living in northern and eastern Europe. Laura Siragusa did her fieldwork with the Veps, a Finno-Ugrian ethnic minority in Russia. Her essay investigates verbal manifestations referring to human–animal relationships that allow deep insights into the ways humans negotiate responsibility for the more-than-human world. In this context, wolves are also seen as symbols for neglected rural areas.

Whether a cohabitation of humans and wolves in the same territories is possible at all is the topic of Helena Ruotsala's case study of reindeer herders in Lapland. Lapland is popularly imagined as an area where wild animals enjoy having enough space. But even here, conflicts between reindeer herders, on the one hand, and wild animals and their advocates, on the other, are becoming increasingly pronounced. By analysing these conflicts, the author considers ways in which the differing interests might be reconciled with each other.

Together, the collected contributions offer a variety of insights into a multivoiced and highly ambivalent situation in the face of the dramatic changes on earth at the beginning of the twenty-first century. They document and study a continuous process of learning that is full of ambiguities and demands a lot from both humans and other-than-humans.

Notes

1 See Large Carnivore Initiative for Europe: Wolf. Canis Lupus. Wolf Facts. https:// www.lcie.org/Largecarnivores/Wolf.aspx [accessed 31 December 2019].
2 In addition to this volume, see the publication of the international conference "Encounters with Wolves. Dynamics and Futures" in Bautzen, June 2018, organised by the editors and some contributors of this volume: Heyer and Hose, in press.

Bibliography

Aisher, A. and V. Damodaran. 2016. "Introduction: Human–Nature Interactions Through a Multispecies Lens." *Conservation and Society* 14 (4), 293–304.

Altex, ed. 2019. Wölfen begegnen. *Tierethik. Zeitschrift zur Mensch-Tier-Beziehung* 11 (19).

Borgards, R. and N. Pethes, eds. 2013. *Tier – Experiment – Literatur 1880–2010*. Würzburg: Königshausen & Neumann.

Bovenkerk, B. and J. Keulartz, eds. 2016. *Animal Ethics in the Age of Humans: Blurring Boundaries in Human–Animal Relationships*. Cham: Springer.

Braverman, I. 2015. *Wild Life: The Institution of Nature*. Stanford, CA: Stanford University Press.

Butler, J. 1993. *Bodies that Matter. On the Discursive Limits of "Sex"*. London and New York, NY: Routledge.

Fenske, M. 2018. "Wolves Eating Sandwiches with Jam: Narrations in Times of New Encounters with the Wild." Paper presented at the Meeting of the International Society for Folk Narrative Research, Ragusa.

Fenske, M. and M. Heyer. 2019. "Wer zum Haushalt gehört: Ethiken des Zusammenlebens in der Diskussion." *Tierethik* 11 (19), 12–33.

Haraway, D. 2008. *When Species Meet*. Minneapolis, MN: University of Minnesota Press.

Heyer, M. and S. Hose. In press. *Encounters with Wolves*. Bautzen, Germany: Sorbian Institute.

Hill, C. M. 2017. "Introduction. Complex Problems: Using a Biosocial Approach to Understanding Human–Wildlife Interactions." In *Understanding Conflicts About Wildlife: A Biosocial Approach*, edited by C. M. Hill, A. D. Webber, and N. E. C. Priston, 1–14. New York, NY: Berghahn.

Imhoof, M. 2012. "More than Honey." Film. Berlin: Senator.

Kirksey, E. 2015. *Emergent Ecologies*. Durham, NC and London: Duke University Press.

Kirksey, S. E. and S. Helmreich. 2010. "The Emergence of Multispecies Ethnography." *Cultural Anthropology* 25 (4), 545–576.

Knight, J. 2000. "Introduction." In *Natural Enemies: People–Wildlife Conflicts in Anthropological Perspective*, edited by J. Knight, 1–35. London and New York, NY: Routledge.

Lorimer, J. 2015. *Wildlife in the Anthropocene: Conservation after Nature*. Minneapolis, MN and London: University of Minnesota Press.

Lorimer, J. and C. Driessen. 2014. "Wild Experiments at the Oostvaardersplassen: Rethinking Environmentalism in the Anthropocene." *Transactions of the Institute of British Geographers* 39 (2), 169–181.

Marvin, G. and S. McHugh, eds. 2019. *Routledge Handbook of Human–Animal Studies*. London and New York, NY: Routledge.

Molinari, G. 2018. *Hier ist noch alles möglich*. Berlin: Aufbau.

Monbiot, G. 2017. "Insectageddon: The Scale and Speed of Environmental Collapse is beyond Imagination." *The Guardian*, 20 October 2017. http://www.monbiot.com/2017/10/23/insectageddon/ [accessed 18 February 2018].

Multispecies Editing Collective, The, ed. 2017. "Troubling Species: Care and Belonging in a Relational World." *RCC Perspectives: Transformations in Environment and Society* 1.

Nevin, O., I. Convery, and P. Davis. (2019). *The Bear: Culture, Nature, Heritage*. Woodbridge, UK: The Boydell Press.

Rosa, H. 2016. *Resonanz. Eine Soziologie der Weltbeziehung*. Berlin: Suhrkamp.

Schimmelpfennig, R. 2016. *An einem klaren, eiskalten Januarmorgen zu Beginn des 21. Jahrhunderts*. Frankfurt am Main: S. Fischer.

Shojaei Kawan, C. 2003. "Rotkäppchen." In *Enzyklopädie des Märchens 11. Handwörterbuch zur historischen und vergleichenden Erzählforschung*, edited by R. W. Brednich, H. Alzheimer, H. Bausinger, W. Brückner, D. Drascek, H. Gerndt, I. Köhler-Zülch, K. Roth, and H.-J. Uther, 854–868. Berlin and Boston, MA: De Gruyter.

Stanišić, S. 2016. *Fallensteller*. München: Luchterhand.

Stengers, I. 2015. *In Catastrophic Times: Resisting the Common Barbarism*. Paris: Open Humanities Press/Meson Press.

van Dooren, T., E. Kirksey, and U. Münster. 2016. "Multispecies Studies: Cultivating Arts of Attentiveness." *Environmental Humanities* 8 (1), 1–23.

Whatmore, S. 2019. "Irdische Kräfte und affective Umwelten. Eine ontologische Hochwasserpolitik." In *NaturenKulturen; Denkräume und Werkzeuge für neue politische Ökologien*, edited by F. Gesing, M. Knecht, M. Flitner, and K. Amelang, 83–104. Bielefeld: transcript.

Woods, M. 2017. *Rethinking Wilderness*. Peterborough: Broadview Press.

Wright, K. and C. Simpson. 2014. "Rethinking Anthropology in the Anthropocene: Knowledges, Practices, Ethics and Politics." *Journal of Media Arts Culture* 11 (1). http://scan.net.au/scn/journal/vol11number1/Simpson-Wright.html [accessed 23 April 2016].

Zapf, H., ed. 2016. *Handbook of Ecocriticism and Cultural Ecology*. Berlin: de Gruyter.

2 The Beast of Gévaudan as a history of the changing perceptions of fatal human–wolf interaction

Meret Fehlmann

The Beast of Gévaudan was a wolf-like animal that terrorized the historical French region of the same name between 1764 and 1767.[1] The attacks of the Beast of Gévaudan resulted in nearly 100 victims. Until the early twentieth century, wolves were generally considered as the perpetrators of these attacks. In this perspective, the Beast of Gévaudan served as a prime example for the threats caused by wolves. The history of the Beast of Gévaudan still fascinates the public, as the events and the nature of the Beast(s) have never been fully enlightened. Speculations and rumours are widespread, and they are constantly enriched by ideas that reflect on preoccupations of the respective time regarding wolves. An important shift in the perception of the Beast of Gévaudan can be linked to the (near) extinction of wolves in Europe and their subsequent protection as an endangered species. In the attempt to exonerate wolves of attacking humans, the Beast of Gévaudan is no longer considered a wolf but is reimagined or reinstated as an exotic and/or domesticated animal, having recourse to the fashionable idea since 1900 that the mind behind the Beast of Gévaudan was a sadistic criminal.

This chapter presents the on-going discussions about the nature of the Beast of Gévaudan as part of the many-voiced discourse concerning the return of wolves in Europe. Its focus lies on popular nonfiction books (from the 1970s onwards) and their changing perception of the Beast that can be inscribed in the contested discourse about wolf-threatening and wolf management strategies. It ends with a quick glimpse into the Beast's appropriation in the regional tourism, thus, bearing witness to its integration into French memorial culture.

The historical events

The first known victim of the Beast of Gévaudan was the 14-year-old Jeanne Boulet, who was attacked and killed on 30 June 1764. Before that date, there had already been a few attacks (maybe even deaths), but they were not attributed to the *bête féroce* – the wild, ferocious beast (Crouzet 1987, 7; Moriceau 2007a, 177). Most of the victims of the Beast were young cattle herders. That children and adolescents were among the favourite victims of the animal was due to the social and economic structure of the region. Gévaudan in the mid-eighteenth century was a poor, underdeveloped, sparsely populated region with patchwork hamlets, and small-scale cattle breeding was common. Young children – alone or in small groups – looked after the cattle near or in the woods. This tradition was partly responsible for the high numbers of assaults (Alleau and Linnell 2015, 97; Buffière 1987, 22; Crouzet 1987, 34; Dixon 2013, 239; Moriceau 2009, 21f.).

As the attacks continued all summer,[2] in the autumn of 1764 King Louis XV stationed a regiment of 57 dragoons nearby. Their wolf management strategy consisted mainly of *battues*,[3] traps and poisoned bait – wolves were killed, but the attacks continued.

The dragoons were withdrawn and replaced by the renowned Norman wolf hunters named D'Enneval – father and son. They continued with the *battues*, with no more success than their predecessors. They were not only unpopular with the inhabitants but also had some altercations with local nobles, so that Louis XV dismissed them and sent his personal *porte-arquebuse* – bearer of weapons – François Antoine de Beauterne, who killed a large wolf near the Abbey of Chazes in September 1765. However, the assaults resumed in the same winter, and now the local populace was left unprotected. They searched for religious comfort and undertook pilgrimages to sanctuaries of the Virgin Mary. It is said that a local hunter had his bullets blessed and killed another animal on the 27 June 1767. Finally, the attacks stopped (Moriceau 2009).

Many rumours concerning the character and nature of the Beast circulated. A contemporary text that offered different explanations for the Beast's apparition was the *mandement* – pastoral letter – of the local bishop Gabriel-Florent de Choiseul Beaupré, declaring that God had sent the Beast as a scourge for the sins of the inhabitants (Crouzet 1987, 8f.; Dixon 2013; Poujade 1985, 31; Smith 2011, 53; Velay-Vallentin 1995, 116–119). The *mandement* also prepared the hypothesis of the exotic animal: "A ferocious Bête, unknown in our climates, has appeared here without knowing where she could have come from" (Choiseul-Beaupré in Pourcher 2006, 60). The idea that the Beast was an exotic animal proved very compelling for the audience – both then and now; the Beast of Gévaudan was not the first man-eater that had been reimagined as an exotic animal – the hyena was an especially popular culprit. A man-eating Beast in the Lyonnais, for example, inspired a *Dissertation sur l'hyène* (Tolomas [1756] 2009). Reports on hyenas dating from the 1750s and 1760s are characterised by a crude mixture of fears, exaggerated stories and scepticism (Poujade 1985, 35; Smith 2016, 36, 48). In 1761, the ninth volume of Georges-Louis Leclerc Comte de Buffon's *Histoire Naturelle* containing an entry on the hyena was published (Buffon 1761).[4] This seminal work with its illustrations was widely read, so the Beast was readily identified as a hyena (Poujade 1985, 35). Buffon eliminated many of the prejudices against the animal; nevertheless, the hyena remains, for him, an aggressive creature of the darkness (Smith 2016, 48, 53). As the historian Jay M. Smith notes: "conventional explanations of rural predation proved less compelling than an animal that stood as an icon of

Bue feroce qui a ravagé le gevaudan en l'anneé 1764

Sion ne doit plus mettre en doute la forme & la figure de l'animal feroce qui ravage le jevaudan, dont on parloit avec si peu de certitude, pour amuser le public, en voici le vrai portrait envoyé a m.' le Prevost de la quatredale avisez par m.' l'abbe Petit de Mendes, qui se trouva a la vue du facheux spectacle de cette jeune fille, qui fui de vorce a la distance de deux coup de fusil, d'un hameau appellé S.' jean des prez ou m.' petit avoit porte le viatique a la tante de la jeune fille, epouse du nommé joseph figuiere, menager, le nombre des habitants fut trop petit pour oser y porter du secours, on prie tous les chretiens d'unir des serventes prieres a celles de M.g.' levêque de mendes, pour la delivrance de ce Monstre, se digne prelat prend les mesures possible pour le detruire, et a ordonné des prieres publiques dans son diocese
M.reys.

Figure 2.1 Historical print (1764) showing the Beast of Gévaudan attacking a woman

mystery – the African hyena" (Smith 2016, 36). During the eighteenth century, due to colonial expansion, exotic animals had become increasingly well-known, so further candidates for the Beast of Gévaudan are the African wild dog, great apes, big cats or even extinct animals.

The image of the wolf and the Beast of Gévaudan

The image of the wolf is burdened by negative stereotypes: the Bible, fairy tales, legends and natural histories all had their share in this development (Campion-Vincent 2005, 99). Buffon's treatise on the wolf in his *Histoire Naturelle* is no exception. His evaluation of the wolf is devastating: "Enfin, desagréable en tout, la mine basse, l'aspect sauvage, la voix effrayante, l'odeur insupportable, le naturel pervers, les mœurs féroces, il est odieux, nuisible de son vivant, inutile après sa mort" (Buffon 1749, 52).[5] The impact of his assessment is still perceptible.

There is an on-going argument whether wolves pose a risk to humans. Historians emphasise the danger of wolves for humans, while zoologists, ethologists and ecologists stress rather that wolves do not (normally) attack humans (Colin 1990, 5; Moriceau 2007a). Thus, no wonder that the Beast of Gévaudan and the question of its nature became a prime target in this debate. The French historian Jean-Marc Moriceau has shown that wolves killed over 3000 people in France from the fifteenth to the nineteenth century.[6] As his work has pointed out, a few single wolves perpetrated most of the attacks. When a wolf was predating on humans, nearly 10 per cent of the annual

Figure 2.2 The Beast of Gévaudan imagined as hyena; historical print of 1765

deaths could be attributed to it. Such a high number surely traumatized the local people, who tried to make sense of a wolf that had left the boundaries of what was taken for a normal lupine behaviour behind and had become something harmful (Moriceau 2014, 226–229).

The Beast of Gévaudan was neither the first nor the last Beast in France,[7] but the 1760s were a time of increasing media coverage. During the lifetime of the Beast, leaflets, journals, gazettes and almanacs reported its misdeeds, changing it into a Europe-wide sinister 'celebrity'. Its transformation from a wild, dangerous animal into a metaphysical monster was also emphasised by reports in the media (Poujade 1985, 51; Séité 1992, 145–149).

The Beast started to inspire novelizations in the early-nineteenth century, and it was only a matter of time until the Beast was reappropriated as an object of scientific interest. The first example of this new take on the Beast is Abbé Pierre Pourcher's *Histoire de la Bête du Gévaudan. Véritable fléau de Dieu* (1889). The subtitle, *True Scourge of God*, indicates that Pourcher understood the Beast of Gévaudan as a metaphysical entity sent by God.[8] Abbé François Fabre ([1930] 1999) argued in his studies that the deeds of the Beast should be attributed to wolves, an explanation that found advocates among many historians (Richard 1999, n.p.) and was generally accepted until the wolf's image underwent some fundamental changes in the 1980s. The 'new wolf' has lost the negative attributes. It has been rediscovered as socially intelligent and serves as an icon of intact wilderness and ecological riches. Furthermore, wolves can be employed as an embodiment of instincts and a

Figure 2.3 German historical print reporting on the misdeeds of the Beast

symbol of a harmonious way of living (Figari and Skogen 2011, 319–322; Moriceau 2007a: 14). Clarissa Pinkola-Estes's bestseller *Women Who Run with the Wolves. Myths and Stories of the Wild Woman Archetype* (1992) is a prime example of this new understanding of wolves and wildlife as symbols of the supressed interior life of modern humanity.

In the case of the Beast of Gévaudan, this new image of the wolf resulted in an attempt at exoneration. There is one text from the early twentieth century that helped to promote a new explanation of the events of the 1760s: *Qu'était la Bête du Gévaudan?* (What Was the Beast of Gévaudan?) by Dr Puech (1911). For him, the Beast is the result of the scared and frenzied imagination and was not a real animal. The victims should be attributed to the crimes of a *fou sadique* – a sadistic madman.

He argues as a modern, enlightened and educated man, referring to the then new sciences, such as the psychology of the masses and the newly sparked interest in serial killers. In 1897/98, France was in the thrall of Joseph Vacher, nicknamed *Le Tueur de Bergers* – killer of shepherds. Among Vacher's victims were several young shepherds, who he killed and raped (Smith 2011, 265–267). This criminal case certainly inspired Puech's vision of the *fou sadique*. The hypothesis of a sadistic madman behind the attacks and not a wild animal has gained in popularity since the 1930s and 1940s. As a result, fiction and nonfiction books that do not rely on the *fou sadique* as the

Figure 2.4 Double statue of the Beast as a *fou sadique* and his trained dog, Le Malzieu

perpetrator are rare nowadays. The main merit of this development can be attributed to the novels *La Bête du Gévaudan* by Abel Chevalley (1936) and *Histoire fidèle de la Bête en Gévaudan* by Henri Pourrat (1946). Pourrat was heavily influenced by Chevalley's retelling; both novels have in common that they present the Beast as the weapon of a group of sadists and that they are often read not as novels but as documentaries of the events of the 1760s (Buffière 1987, 170). Both authors lead the readers astray: Chevalley by using a fictional editor telling the story of how he found his grandfather's notes that reveal the truth. Pourrat calls his novel simply the *true, faithful history – histoire fidèle –* thus, implying that the truth can be found in his work.

The Beast of Gévaudan and the protection of wolves

Wolves were nearly extinct in France around 1900 and were finally put under protection in 1979 (Ragache and Ragache 1981, 210–213; Smith 2011, 267). However, there were 39 wolf sightings between 1945 and 1989 of which 29 ended with the death of the wolf, thus, confirming the existence of wolves in France. Those wolf sightings aroused suspicion and it was rumoured that the wolves did not return naturally but had escaped from wildlife parks or had even been reintroduced by ecologists (Campion-Vincent 2002, 30f.; 2005, 112).

In the region that was once the Gévaudan, there lived supposedly some 150 to 200 wolves after World War I. Since the 1980s, there have been rumours that the wolves have returned to the Margeride Mountains – the same region where the Beast was killed. However, the reappearance in this area was only affirmed as late as 2005, while the first confirmed comeback of wolves in France was in the early 1990s, coming from Italy (Campion-Vincent 2005, 101; Soulier 1995, 136).

An interesting work concerning the disappearance of wolves is Jacques Delperrié de Bayac's (1970) *Du sang dans la montagne. Vrais et faux mystères de la Bête du Gévaudan* (Blood in the Mountains. True and False Mysteries of the Beast of Gévaudan), published nearly a decade before wolves became a protected species and two decades before their actual return to France. It is a nonfiction book retracing the history of the Beast of Gévaudan. Delperrié de Bayac distinguishes between fact and fiction and notes that Puech's hypothesis of the *fou sadique* fits the taste of a modern audience and has influenced many retellings, such as Abel Chevalley's *La Bête du Gévaudan* and Henri Pourrat's *Histoire fidèle de la Bête en Gévaudan* (Delperrié de Bayac 1970, 250–252). He is of the opinion that several wolves perpetrated the deeds of the Beast. In his evaluation, the wolves were responsible for many carnages in French history, a fact that is reflected in the book cover showing a wolf with bared fangs. This is an interesting or even contradictory iconography for a book that laments the extinction of the species as a tragedy:

En fait, même si l'espèce n'a pas totalement disparu de France, le loup, jadis premier brigand de nos campagnes, a déjà, chez nous, rejoint sa légende, et comme j'arrive à la fin de ce livre où j'ai raconté tant de meurtres et de carnages, je dois dire maintenant le plus difficile: le loup n'était pas que malfaisant, il avait son utilité et son rôle, il contribuait à l'équilibre des espèces.

(Delperrié de Bayac 1970, 265)[9]

His affirmation that the wolf was not only evil but had its utility, is a response to Buffon's famous maxim that the wolf is harmful in life and useless in death. Delperrié de Bayac mentions the passing of the wolf into the realm of legend. During the twentieth century, the wolf became a legendary being in large parts of Europe, first considered with fear and suspicion, and then changed into a symbol of intact wilderness. Discourses about wolves have been oscillating between these two viewpoints up until today.

Delperrié de Bayac mentions the alleged wolf sightings in France since the end of World War II. He was convinced that there were still some wolves or hybrids of wolf and dog. However, as the wolf has reputedly been extinct since the early twentieth century, he is persuaded that farmers and hunters can no longer tell the tracks of wolves and dogs apart. He reveals in his book that he is knowledgeable about wolves, having kept several (see pictures in Delperrié de Bayac 1970).

He ends his long book about the murderous actions of the Beast of Gévaudan and wolves in the history of France with a complaint about the French delay regarding the protection of the environment. The Beast functions as a symbol of an unenlightened time:

Nous ne sommes plus au temps de la Bête du Gévaudan. Le très grand retard pris par notre pays dans le domaine de la protection de la faune et de la flore, l'absence d'une véritable politique de la nature, des habitudes (mauvaises), des intérêts (très puissants), n'ont pas permis à la France d'en faire autant. On peut penser que c'est dommage. Je le pense.

(Delperrié de Bayac 1970, 266)[10]

Delperrié de Bayac remains a mystery; he was a historian, author and journalist (Bibliothèque nationale de France 2017). He was politically left-wing and an early ecologist (Campion-Vincent 2002, 28f.; Greisalmer 1994, n.p.; Ragache and Ragache 1981, 250).[11] Related to the rumours that the wolves did not return naturally but were reintroduced, his name is sometimes mentioned in connection with the case of the wolves from the Landes (Southwest France) in 1968. As the wolves were a couple, it was often thought that they were deliberately set free. It is said or written that he bought a couple of wolves from Gérard Ménatory, the founder of the *parc à loups du Gévaudan*, and released them in the Landes. However, this experiment was only short-lived, because the wolves, not accustomed to a life of liberty, did not last long in their new, rather hostile environment (Campion-Vincent 1992, 162; 2002, 28; Delperrié de Bayac 1970, 263f.).

Ménatory has written many books about wolves and the Beast of Gévaudan in the hope of exonerating the species and helping people to overcome their fear of wolves. He published *La Bête du Gévaudan* in 1976. Like Delperrié de Bayac a few years earlier, Ménatory stresses the importance of the protection of nature at the end of his book, ending with a plea for the acceptance of wolves that cannot defend themselves:

> Je ne dois pas oublier, par ailleurs (je n'ai pas le droit de l'oublier) que je suis un protecteur de la Nature. Or, en remettant un peu les choses à leur place, dans cette sombre histoire, je dégage les loups de très lourdes responsabilités. Responsabilités qu'ils n'ont pas à endosser; mais ils n'avaient pas, eux, les moyens de clamer leur innocence, et ce fut bien commode pour certains.
>
> (Ménatory 1976, 129f.)[12]

For Ménatory, the stories of man-eating Beasts must be regarded as legends and the expression of the irrational fear of wolves of a population who must no longer deal with them – a message he has repeated many times. This argument laid the foundation for many to concur that the wolves were innocent of killing the 100 victims of the so-called Beast of Gévaudan. His denouement sees as a culprit a hyena trained and abused by humans for their sinister ends. It is a story of human manipulation of animals that serves to exonerate the (wild) animal. This narrative pattern has gained enormously in popularity since the 1970s (Campion-Vincent 1992, 163). In the same decade, Ménatory was probably the first author to rediscover the hyena (Smith 2016, 34). The hyena is already an old acquaintance, having gained some notoriety as a possible culprit in the 1760s and in the novels of Chevalley and Pourrat in the first half of the twentieth century. Both authors, but especially Chevalley, served Ménatory as inspiration (Ménatory 1976, 123f.).

The innocence of wolves

As wolves were (nearly) extinct in twentieth-century France, knowledge about their life and behaviour was sparse. Their absence in real life did not hinder that collective beliefs about wolves were full of negative bias. However, since the 1980s, the image of the wolf has undergone a dramatic change towards the positive (Soulier 1995, 130). As noted earlier, the innocence of wolves called for placing the blame on new culprits in the history of the Beast of Gévaudan. A new culprit was found in the *fou sadique* – a figure that has left behind his humanity by abusing animal(s) to kill children and women. This idea illustrates the 'violence graduation hypothesis' often mentioned in literature on criminal behaviour and popular opinion, whereas animal abuse often escalates over time to acts that harm people (cf. Arluke et al. 1999). What seems like a modern idea has its roots in the animal rights movement of the nineteenth century that advocated the notion that friendship with animals should serve as a means of refinement of character. Consequently, neglecting or abusing animals indicates the moral decline of the individual (Buchner-Fuhs 1999, 275–277). This belief can be found in Ménatory's *La Bête du Gévaudan*, where he suggests that a hyena was trained to attack to gratify the sadistic tendencies of its owner, because killing for pleasure is characteristic only for humans (Ménatory 1976, 109).

One publication that already emphasises the innocence of wolves in its title is Louis' *La Bête du Gévaudan. L'innocence des loups* (The Beast of Gévaudan. The Innocence of Wolves), which was first published in 1992, chronologically in accordance with the wolves' return to France. His main statement concerning the Beast is as follows: "C'est une MACHINE À TUER qui résulte d'un DRESSAGE méné par une main diabolique et criminelle" (Louis 1997 239).[13] Here, the culprit behind the Beast is man, who changed an innocent animal into a killer. Louis wants to prove the innocence of wolves and seems to have a background in biology. However, his argument tends to mix up all kind of literature, thus confusing fact and fiction when he reads, for example, the novels of Chevalley and Pourrat at face value to support the idea that the misdeeds of the Beast of Gévaudan can be attributed to the plots of sadistic men.

One of the fervent advocates of the innocence of wolves is Hervé Boyac's (2013) in his *La Bête du Gévaudan. Le loup acquitté, enfin!* (The Beast of Gévaudan. The Wolf Finally Acquitted!). The inside cover of the book contains some information on the author: "Membre actif de Ferus qui défend les grands prédateurs, il milite pour un retour accepté du loup dans notre pays" (Boyac 2013, n.p.).[14] *Ferus* (wild, savage in Latin) is a group that advocates the return of the great predators in France and the coexistence of wolves and pastoralism (Ferus 2017); it was formed in 2003 by merging the *Group Loup France* and *ARTUS* (a group for the preservation of bears; Campion-Vincent 2005, 101).

For Boyac, the Beast was not a wolf but an exotic animal trained to attack humans. He envisions big cats, wolf–dog hybrids or the old hyena as possible candidates. The real culprit is man, with his tendency to abuse animals (Boyac 2013, 403). His book ends with a plea for the acceptance of wolves even surpassing the aims of *Ferus:* "Il faut savoir que le loup, présent depuis la préhistoire sur notre sol, est plus naturel que le mouton qui a été obtenu à partir du mouflon importé du Moyen-Orient voilà 8000 à 10000 ans seulement, au moment de la sédentarisation de l'homme" (Boyac 2013, 502).[15]

His argument that the wolf is more natural in Europe than sheep and goats, animals that came into being through human domestication, indicates one of the important points of dissent among supporters and adversaries of the return of wolves: What is nature? Is it wilderness or is it land cultivated and formed by human hands? In this dichotomy, the wolf is the messenger of wilderness and stands for the absence of human influence, whereas sheep and goats embody the human effort to shape the landscape. The return of the wolf in the South of France is sometimes seen as a further step in the direction of the general neglect of the region that is affected by depopulation – the younger generation moves to the big cities, for example, Lyon and Paris – so the wolf is not only the symbol of wilderness but also becomes a danger to the human efforts to cultivate and (re-)form the landscape. Therefore, the wolves' return to France can be interpreted as wilderness growing out of control (Campion-Vincent 2005, 116; Lescureux 2001/2, 14–17).

Furthermore, Boyac sees a close relationship between the wolf and original humanity, which has been disturbed by layers of civilization: "Pour comprendre et accepter le loup, nous devons expurger nos frustrations internes 'd'hommes modernes', qui nous dirigent et nous empêchent d'avoir une réflexion pragmatique sur notre place et notre rôle sur la planète terre" (Boyac 2013, 474).[16] The reference to the "frustrations of modern man" recalls, in my opinion, the use of the image of the wolf of Clarissa Pinkola-Estes's (1992) bestseller *Women who Run with Wolves. Myths and Stories of the Wild Woman Archetype*. One of her central points is the equation of ecocide and the modern psychosocial condition: "It's not by accident that the pristine wilderness of our planet disappears as the understanding of our own inner wild nature fades" (Pinkola-Estes 1992, 3). One evident difference is that her book addresses solely women, stressing the similarities of wilderness, wolves and women and their oppression by environmental pollution, reclining spaces of natural wilderness and a patriarchal society. For her, the wolf symbolizes liberated femininity in line with a regained sense of the importance of the (feminine) instincts.

Tourism and the Beast of Gévaudan

Nowadays, the *départements* of Lozère and Haute Loire are – as in the days of the Beast of Gévaudan – rather under-developed, isolated, provincial regions (Fournier 2001, 135). There is not much tourism; people drive through to get to the more glamorous South of France. Black Madonnas, impressive Romanesque churches and the "Way of St. James" attract spiritually interested tourists but otherwise there is not much to promote tourism except for nature and the Beast of Gévaudan. Thus, it does not come as a surprise that the Beast plays an important role in promoting the tourism of the region. As Laurent Fournier notes in his study *Petite histoire des grands ravages d'une méchante bête* [Short History of the Big Devastation of a Mean Beast], the Beast of Gévaudan is the biggest attraction the region has ever known (Fournier 2001, 128). The connection to the Beast is employed in the hope of attracting people and, ultimately, financial gain. Local shopkeepers allude to the Beast in their marketing; a baker has an eponymous speciality and his wrapping paper announces:

> Elle a mangé nos ancêtres
> Les miens, les vôtres, peut-être
> Vengez-vous donc en croquant
> La bête du Gévaudan!

(Soulier 1995, 135)[17]

For many locals, the reference to the Beast is linked to the hope of a better economic future. Local Beast-enthusiast Jean Richard writes in line with this view: "[...] son retour peut apporter espoir et mieux-vivre. [...] Et qui sait si, elle qui tua une centaine de personnes, en 1764–67, ne pourrait pas permettre à autant de gens de (re)vivre et travailler au pays à l'aube de l'an 2000" (Richard 2015, 40f).[18] However, the opinions concerning the beneficial effects of the Beast are divided. There are those who criticize this use of the region's past as superficial cosmetics, at its best: "L'alibi touristique a donc pour fonction essentielle de masquer le sous-développement et la colonisation administrative et économique du département, tout en donnant l'illusion d'un espoir de survie. Une nouvelle Bête du Gévaudan pour les indigènes?" (Fournier 2001, 136).[19] Fournier criticizes the strong French centralism, with Paris as the centre, leaving the regions powerless. He cannot see a beneficial effect in the tourism but rather a repetition of the past that left the locals with the harm.

Nonetheless, the Beast of Gévaudan and its memories have inspired some interesting attractions for tourists. Gérard Ménatory founded his *parc à loup du Gévaudan*, a tourism hotspot, in this region. An early enthusiast of wolves, he went to the Białowieża National Park in 1961 where he was able to procure two whelps, which were the foundation for his originally private animal park (Soulier 1995, 133). In the 1980s, he merged his wolves with the pack of another wolf park; thus, *le parc à loup du Gévaudan* came into being (Parc à loup du Gévaudan 2017a). Since the early 1990s, the wolf park has attracted more than 100,000 visitors yearly who appreciate seeing wolves up close. When visiting captive wolves, the visitor expects a kind of substitute for the wild wolf. The *Group loup France* (the precursor of *Ferus*) considered *le parc à loup du Gévaudan* a commercial offer, but, nevertheless, decided that it was better having wildlife parks than no possibility of seeing wolves at all. So these parks are believed to play an important role in raising broader public awareness for the needs of the wild wolves (Chabert et al. 2004, 134).

As the *parc à loup du Gévaudan* wants to improve the general acceptance of wolves, its homepage contains some information about the Beast of Gévaudan. One part of it is dedicated to Ménatory's hypothesis of the trained hyena (Parc à loup du Gévaudan 2017b). This interpretation is quite fitting for an institution that wants to promote a new understanding of wolves as important players in the balance of nature. The visual material of the homepage reflects the effort to present wolves in the best possible light. They are shown in peaceful poses, often in groups, thus, stressing the new image of the wolf as a socially intelligent animal (Figari and Skogen 2011, 319–322; Moriceau 2007a, 14).

Since 1999, the Beast has even had its own *musée fantastique de la Bête du Gévaudan* (Fantastic Museum of the Beast of Gévaudan) in the small town of Saugues. This *son-et-lumière* spectacle retraces the life and death of the infamous animal in 22 scenes with about 60 life-size figures. The museum is the result of the efforts of the local artist and Beast-enthusiast Lucien Girès, who created it with the help of the association Macbet (Musée fantastique de la Bête du Gévaudan, Association Macbet 2017). Girès' pictures and cartoons of the Beast can be seen wherever one goes in the region; the so-called *route de la Bête* that connects Saint-Chély D'Apcher and Le Puy has its own traffic panel complete with a Beast standing on its hind legs by Girès (Richard 2015, 55). In this case, the reference to the Beast has nothing to do with respect for wolves but is an attempt to profit from the local history by attracting passing tourists.

Since the mid-twentieth century, nearly every town and village has had its own statue of the Beast (Buffière 1987, 26; Soulier 1995, 135). Those statues tell quite different stories of the Beast. It can appear in a naturalistic manner as a wolf or wild animal, or it can be rendered as a *fou sadique* accompanied by his trained animal – even in the same village – thus, bearing witness to the longevity of the Beast of Gévaudan whose mysteries have been adapted to fit many needs.

Final remarks

Instead of a classic conclusion, one could state that the Beast of Gévaudan is quite alive in the region itself and lives on comfortably in books of all genres.

The last decades have seen a new interpretation of the events of the 1760s that tries to relieve wolves of all involvement with the attacks. Instead, wolves are portrayed as wild and noble animals. This tendency manifests itself not only in nonfiction books, such as the examples I have considered here, but also in novelizations that often portray wolves as protectors against the Beast of Gévaudan.[20]

Figure 2.5 Statues of the Beast of Gévaudan as a wild animal in Le Malzieu

Whatever the animal was, its misdeeds still hold a thrall over audiences near and far; and now let me end with a quote by Jean Richard: "Une chose est sûre, nous n'avons pas fini d'entendre parler d'elle [...]" (Richard 1999, n.p.).[21]

Notes

1 The historical province of Gévaudan is identical to the modern *départements* of Haute Loire and Lozère.
2 As the historian Moriceau has shown, the season marked by the largest numbers of attacks was the summer (May to mid-September; Moriceau 2014, 218).
3 A method of hunting whereby the game is driven towards the hunters by teams of beaters.
4 Also as a digital resource, http://www.buffon.cnrs.fr/ice/ice_page_detail.php?lang= fr&type=text&bdd=buffon&table=buffon_hn&typeofbookDes=hn&bookId=9&pa geChapter=L%E2%80%99Hy%C3%A6ne.&pageOrder=270&facsimile=off&search =no&num=&nav=1 [accessed 10 April 2017].
5 "Finally, obnoxious in every aspect, of low facial expression, wild appearance, frightening voice, unbearable smell, perverse nature, vicious morals, he is hateful, harmful in life and useless in death."
6 In 2014, he augmented the count to 9000 victims of wolves in France alone.
7 The Beast of the Gâtinais (1652–1657), the Beast of the Lyonnais (1754–1756) or the Beast of the Cévennes (1809–1816) were also notable (cf. Alleau 2013, 55f.).
8 He collected the oral lore concerning the Beast and archival records for his extensive work – more than 1000 pages. This work remains a rich source of lore about the Beast. The book was translated into English by Derek Brockis as *The Beast of Gevaudan* in 2006. Since the early twenty-first century, the Beast has left its French context and has manifested in different languages and cultures. This is probably due to the box office success of the French movie *Le pacte des loups* (Christoph Gans, 2001) which was released in the USA as *Brotherhood of the Wolf*.
9 "Even if the species hasn't totally disappeared from France, the wolf, once first bandit of our countryside, has already rejoined his legend, and as I come to the end of this book where I have told about so many murders and massacres, I have now to say the most difficult thing: the wolf wasn't just evil, he had his utility and his role, he contributed to the equilibrium of the species".
10 "We no longer live in the period of the Beast of Gévaudan. The very big delay of our country in the field of the protection of wildlife and plants, the absence of an actual policy of the environment, the (bad) habits, the (strong) interests haven't allowed France to do as much. You can think this is a pity. I think so".
11 He has written since the 1960s about the popular front in France, the international brigades in Spain and wolves. See, for example, https://fr.wikipedia.org/w/index.php? title=Jacques_Delperri%C3%A9_de_Bayac&oldid=124008513 [accessed 10 April 2017]. Even his entry in the *Bibliothèque Nationale de France* comprises only his date of birth: http://catalogue.bnf.fr/ark:/12148/cb12370628f [accessed 29 April 2017].
12 "I must not forget, furthermore (I don't have the right to forget), I'm a protector of nature. By putting the things back in their place in this bleak history, I free the wolves of heavy responsibilities. Responsibilities they don't have to endorse, but they didn't have the means to proclaim their innocence, and this was convenient for some people".
13 "It's a killing machine, the result of training conducted by a diabolical and criminal hand."
14 Active member of *Ferus* defending the great predators, it/he (*il* can be translated by both pronouns) is militantly in favour of an accepted return of the wolf in our country.

15 "It must be known that the wolf has been present since prehistoric times on our soil, so he is more innate than the sheep that were obtained from the mouflon in the Middle East only 8,000 to 10,000 years ago at the period of sedentarization of man".

16 To understand and accept the wolf, we need to expurgate our frustrations of "modern man" that direct us and prevent us from having pragmatic thoughts concerning our place and role on the planet Earth.

17 "She has eaten our ancestors/Mine, maybe yours/Take revenge by eating/The Beast of Gévaudan!"

18 "Its return can bring hope and better living [...] and who knows, the Beast that killed around hundred people between 1764 and 1767 could allow as many people to live and work in the region at the dawn of the year 2000".

19 "The alibi of tourism has as an essential function to cover up the under-development and the administrative and economic colonization and give the illusion of the hope of survival. A new Beast of Gévaudan for the natives?"

20 A few examples are Pelot (2001), Raven (2008), Blazon (2012) and Hédelin (2014).

21 "One thing is for sure, we haven't finished hearing talk about her [...]".

Bibliography

Alleau, J. 2013. "À qui peut-on faire croire ce genre de sornettes?" *Sens-Dessous* 12 (2), 51–62. http://www.cairn.info/revue-sens-dessous-2013-2-page-51.html [accessed 24 April 2017].

Alleau, J. and J. D. C. Linnell. 2015. "The Story of a Man-eating Beast in Dauphiné, France (1746–1756)." In *A Fairytale in Question. Historical Interactions Between Humans and Wolves*, edited by P. Masius and J. Sprenger, 79–100. Cambridge: The White House Press.

Arluke, A., J. Levin, C. Luke, and F. Ascione. 1999. "The Relationship of Animal Abuse to Violence and Other Forms of Antisocial Behavior." *Journal of Interpersonal Violence* 14, 963–975.

Bibliothèque nationale de France2017. *Delperrié de Bayac, Jacques.* http://catalogue.bnf.fr/ark:/12148/cb12370628f [accessed 29 April 2017].

Blazon, N. 2012. *Wolfszeit*. Ravensburg, Germany: Ravensburger Verlag.

Boyac, H. 2013. *La Bête du Gévaudan. Le loup acquitté, enfin!* Clermont Ferrand: De Borée.

Buchner-Fuhs, J. 1999. "Das Tier als Freund. Überlegungen zur Gefühlsgeschichte im 19. Jahrhundert." In *Tiere und Menschen. Geschichte und Aktualität eines prekären Verhältnisses*, edited by P. Münch, 275–294. Paderborn: Schöningh.

Buffière, F. 1987. *La Bête du Gévaudan. Une (grande) énigme de l'histoire.* Toulouse: Buffière.

Buffon, G. L. 1749. "Le loup." In *Histoire naturelle*, edited by G. L. Buffon, 39–74. Paris: Imprimerie Royale, Vol. 7. http://www.buffon.cnrs.fr/ice/ice_page_detail.php?lang=fr&type=text&bdd=buffon&table=buffon_hn&bookId=7&typeofbookDes=hn&pageChapter=Le+Loup.%0D&pageOrder=52&facsimile=off&search=no [accessed 10 April 2017].

Buffon, G. L. 1761 "L'hyène." In *Histoire naturelle*, edited by G. L. Buffon, Vol. 9, 268–298. Paris: Imprimerie Royale. http://www.buffon.cnrs.fr/ice/ice_page_detail.php?lang=fr&type=text&bdd=buffon&table=buffon_hn&typeofbookDes=hn&bookId=9&pageChapter=L%E2%80%99Hy%C3%A6ne.&pageOrder=270&facsimile=off&search=no&num=&nav=1 [accessed 10 April 2017].

Campion-Vincent, V. 1992. "Appearances of Beasts and Mystery-cats in France." *Folklore* 103, 160–183.

Campion-Vincent, V. 2002. "Les réactions autour du retour du loup en France. Une tentative d'analyse prenant les rumeurs au sérieux." *Le monde alpin et rhodanien. Revue régionale d'ethnologie* no. 1–3, 11–52.

Campion-Vincent, V. 2005. "The Restoration of Wolves in France. Story, Conflicts and Uses of Rumor." In *Mad About Wildlife. Looking at Social Conflict Over Wildlife*, edited by A. Herda-Rapp and T. L. Goedeke, 99–122. Leiden: Brill.

Chabert, J.-P., C. Sainte Marie, and M. Vincent. 2004. "La régularisation du loup 1990–2004." *Forêt méditerranéenne* XXV (2), 131–142.

Chevalley, A. 1936. *La Bête du Gévaudan*. Paris: Gallimard.

Colin, S. 1990. *Autour de la Bête du Gévaudan*. Le Puy en Velay: Imprimerie Jeanne D'Arc.

Crouzet, G. 1987. *Quand sonnait le glas au pays de la Bête*. Clermont-Ferrand. Clermont Ferrand: C.R.D.P.

Delperrié de Bayac, J. 1970. *Du sang dans la montagne. Vrais et faux mystères de la Bête du Gévaudan*. Paris: Fayard.

Dixon, D. 2013. "Wonder, Horror and the Hunt for La Bête in Mid-18th Century France." *Geoforum* XXX, 239–248. http://dx.doi.org/10.1016/j.geoforum.2012.12. 018 [31 March 2017].

Fabre, F. 1999 [1930]. *La Bête du Gévaudan*. Clermont-Ferrand: Editions de Borée.

Ferus (2017). L'association. http://www.ferus.fr/a-propos-de-ferus/association [accessed 4 April 2017].

Figari, H. and K. Skogen. 2011. "Social Representations of the Wolf." *Acta Sociologica* 54 (4), 317–332. doi:10.1177/000169931422090.

Fournier, L. 2001. *Petite histoire des grands ravages d'une méchante bête. Le mystère de la Bête du Gévaudan*. Sémalens: PSR global web.

Greisalmer, L. 1994. "Le retour d'un historien oublié." *Le Monde*, 13 June 1994. https:// www.nexis.com/results/enhdocview.do?docLinkInd=true&ersKey=23_T25790179675 &format=GNBFI&startDocNo=0&resultsUrlKey=0_T25790179677&backKey=20_ T25790179678&csi=299258&docNo=1 [accessed 10 April 2017].

Hédelin, P. 2014. *Les grandes énigmes de l'histoire. La bête du Gévaudan*. Montrouge: Bayard Poche.

Lescureux, N. 2001/2. *Représentation collective du loup dans un village du Mercantour. Les inquiétudes d'une communauté rurale face à son avenir*. Master thesis. Paris: Muséum national de l'histoire naturelle athttp://www.academia.edu/1059023/Repr%C3%A9senta tion_collective_du_loup_dans_un_village_du_Mercantour_les_inqui%C3%A9tudes_ dune_communaut%C3%A9_rurale_face_%C3%A0_son_avenir [accessed 10 April 2017].

Louis, M. 1997 [1992]. *La Bête du Gévaudan. L'innocence des loups*. Paris: Perrin.

Ménatory, G. 1976. *La Bête du Gévaudan. Histoire – légende – réalité*. Mende: Chaptal.

Moriceau, J.-M. 2007a. *Histoire du méchant loup. 3000 attaques sur l'homme en France xv–xx siècle*. Paris: Fayard.

Moriceau, J.-M. 2007b. "Les enfants devorés par les loups dans la France moderne (1590–1820)." In *Histoire des familles, de la démographie et des comportements, en hommage à Jean-Pierre Bardet Poussou*, edited by J.-P. Poussou, 585–593. Paris: PUPS.

Moriceau, J.-M. 2009. *La Bête du Gévaudan*. Paris: Larousse.

Moriceau, J.-M. 2014. "Sous la cruelle dent du loup. 9000 victimes humaines en France du 16ᵉ au 19ᵉ siècle." In *Vivre avec le loup? Trois mille ans de conflit*, edited by J.-M. Moriceau, 211–229. Paris: Tallandier.

Musée fantastique de la Bête du Gévaudan2017. Association Macbet. http://www. musee-bete-gevaudan.com/association-macbet [accessed 8 April 2017].

Parc à loup du Gévaudan2017a. Historique du parc. http://www.loupsdugevaudan. com/decouvrez-le-parc/historique-du-parc/ [accessed 27 April 2017].

Parc à loup du Gévaudan2017b. Version de Gérard Ménatory. http://www.loupsdu gevaudan.com/decouvrez-le-parc/la-bete-du-gevaudan/ [accessed 27 April 2017].

Pelot, P. 2001. *Le pacte des loup.* (A novelization of the film of the same name.) Paris: Rivages.

Pinkola-Estes, C. 1992. *Women Who Run with the Wolves. Myths and Stories of the Wild Woman Archetype.* New York, NY: Ballantine.

Poujade, R. 1985. "La Bête du Gévaudan. Contribution à l'histoire d'un mythe." *Revue du Gévaudan, des Causses et des Cévennes* 1, 25–55.

Pourcher, P. 2006 [1889]. The Beast of Gévaudan. La Bête du Gévaudan. Translated by Dereck Brockis. Bloomington, IN: Author House. (Original work published as *La Bête du Gévaudan. Véritable fléau de Dieu.*)

Pourrat, H. 2011 [1946]. *Histoire fidèle de la Bête en Gévaudan.* Marseille: Lafitte.

Puech, P. 1911. "Qu'était la bête du Gévaudan?" *Mémoires de la Section de médecine/ Académie des sciences et lettres de Montpellier* 2 (4), 409–430. http://gallica.bnf.fr/a rk:/12148/bpt6k441697n/f417.image.r=M%C3%A9moire%20de%20l%27Acad%C3% A9mie%20des%20sciences%20et%20lettres%20de%20Montpellier.langFR [accessed 3 April 2017].

Ragache, C.-C. and G. Ragache. 1981. *Les loups en France. Légendes et réalités.* Paris: Aubier-Montaigne.

Raven, L. 2008. *Werwolf.* Wien: Ueberreuter.

Richard, J. 1999. "Compléments iconographiques, historiques et bibliographiques." In *La Bête du Gévaudan*, edited by F. Fabre. Clermont-Ferrand: Editions de Borée.

Richard, J. 2015 [2013]. *La Bête du Gévaudan dans tous ses états. Dessins – sculptures – peintures – muséographie par Lucien Girès. A l'occasion du 250ème anniversaire de l'apparition de la Bête.* Saugues: Editions des amis de la tour.

Séité, Y. 1992. "La bête du Gévaudan dans les gazettes. Du fait divers à la légende." In *Les gazettes européennes de langue française*, edited by H. Duranton, 145–154. Saint-Etienne: Publications de l'université de Saint-Etienne.

Smith, J. M. 2011. *Monsters of the Gévaudan. The Making of a Beast.* Cambridge, MA: Harvard University Press.

Smith, J. M. 2016. "Dreadful Enemies. The Beast, the Hyena, and Natural History in the Enlightenment." *Modern Intellectual History* 13 (1), 33–61. doi:10.1017/ S14792443150000.

Soulier, B. 1995. "Le loup dans l'imaginaire contemporain du Gévaudan." *Cahiers de l'imaginaire* 14, 121–150.

Tolomas, C.-P.-X. 2009 [1756]. *Dissertation Sur L'Hyene, A L'Occasion De Celle Qui A Paru Dans Le Lyonnois Et Less Provinces Voisoines, Vers Les Derniers Mois De 1754, Pendant 1755 Et 1765.* Whitefish, MO: Kessinger Publishing.

Velay-Vallentin, C. 1995. "Entre fiction et réalité. Le petit chaperon rouge et la bête du Gévaudan." *Gradhiva* 17, 111–126.

3 Made of stone, flesh and narration – 'the wolf' as contested *lieu de mémoire*

Marlis Heyer and Susanne Hose

When a wolf wears sables

Two editorial offices of *Sächsische Zeitung*,[1] a Saxon newspaper, referred to a photographic shoot by local photographer Matthias Schumann in January 2018 to inform their readers about recent statistics concerning the wolf population in Saxony. The rather scurrilous eye-catcher shows the sandstone sculpture of a wolf, a hunting memorial erected at Laußnitzer Heide (Laußnitz Heath) in 1740. Somebody had dressed the wolf in a black hat and scarf. "The lithic animal is wearing sables at the moment",[2] says the caption. Following the journalist's comment, there is "no need to howl"[3]: according to the numbers published by the DBBW,[4] the Saxon wolf population is flourishing. Twelve documented cases of death in 2017, most of them road casualties, are outnumbered by a minimum of 48 newborn wolf pups[5] – not to the delight of everybody (cf. SZ GRH 2018).

This chapter asks how 'the wolf' is discussed in the East German Lusatia region[6] and what role it plays in Lusatian people's collective memory. Media of the collective memory are diverse: orally transmitted and written narratives (legends, fairy tales, belles-lettres, technical literature, encyclopaedias), works of art (paintings, graphics, caricatures, photographs, stamps, heraldry, sculptures), scenic performances (plays, films) and even proverbs, as well as idiomatic expressions and field and personal names tell about human–wolf interactions. How are the images and motifs conveyed renewed in today's narratives when memorized and living wolves meet? When, for example, narrations of the employees of the contact office "Wolfsregion Lausitz", Germany's oldest public institution[7] with the task of mediating between wolves and society, encounter the popular narrative traditions of the bad or foolish wolf in particular or rather the dark forest and the unpredictable nature in general?

The French historian Pierre Nora has introduced the concept of *lieux de mémoire* to describe historic points of reference (events, characters, places, myths) "where memory crystallizes and secretes itself" (Nora and Kritzman 1996, 1). According to this concept, a site of memory is "any significant entity, whether material or non-material in nature, which by dint of human will or the work of time has become a symbolic element of the memorial

Figure 3.1 The wolf wears black
Photographer: Matthias Schumann

heritage of any community" (Nora and Kritzman 1996, XVII). Relating to the historico-cultural research with a national or transnational perspective, memory spaces are usually of importance regarding their identity-giving meaning. For Nora, *lieux de mémoire* were mainly about the identification and analysis of topoi inherent to national memory cultures. Therefore, when the concept was outlined at the end of the 1970s, it was used in a rather narrow and limited frame of national reference (François and Schulze 2001; Nora 1984–94; 2005). Since then, publications such as *European Memory Spaces* (den Boer et al. 2012) and *Ecological Memory Spaces* (Uekötter 2014) have proved the possible broader application areas beyond national historiography. Still, the detection of *lieux de mémoire* serves mostly as an explanation of a – perceived – historically based sense of community and often stuck to place (Siebeck 2017, 9–10). In addition to investigations of shared memories shaping identity and community, the approach of *lieux de mémoire* could also help to analyse shared myths of different kinds. For us as researchers interested in narrative cultures, narrations and narratives belong to these identity-shaping means. In this regard, we ask whether Nora's research design is also applicable to the analysis of the current wolf discourse in East German Lusatia. In other words, is not 'the wolf' itself, in all the different narrative manifestations, a *lieu de mémoire*? It is this question that leads us through popular (and sometimes populistic) debates, following the traces of wolves to their aggregations and knots in a web of mnemotopoi (Kreis 2010, 342).

Lieux de mémoire as an approach to the discourse on 'the wolf' and wolves

Nora's term *lieux de mémoire* provides us with a research paradigm which enables us to investigate not only the narratives themselves but also the surplus of symbolic meaning they produce. *Lieux de mémoire* must be understood as a metaphor in several regards: the term signifies the research design and the objects of investigation. It marks historic events, persons, places and artefacts considered meaningful by human actors, which, therefore, are passed through generations via different cultural techniques. Beyond doubt, narrating is one of the most important mnemonics and possesses a wide arsenal of tools, such as repetitions, schematization, formulization, crystallization, rhythmicity and orchestration (Holbek 1984; Olrik 1909; Parry 1971). All these techniques help to stabilize memories despite processes of social changes and transformations. However, memories are not resistant to change. Contrariwise, they are dynamic and undergo permanent selection and alteration.

Cultural memories of different European societies conserve the wolf in its role as an antagonist of humans with mostly negative connotations (Hiltmann 2013). Despite she-wolves feeding European founding fathers, such as Romulus and Remus, in the canon of traditional narrations, no other animal in the region is as fear-laden. The wolf embodies voracious greed, wickedness and brutality in fables, fairy tales and animal epics (cf. Bies 2014; Rheinheimer 1995). Human actors respond to these characteristics with tantamount cruelty: "Towards the wild wolf, as it happened to the fairy tale wolves of the Brothers Grimm, all desires for torture and slaughter could be acted out"[8] (Schenda 1995, 395). 'The wolf' symbolizes the untamed and unpredictable, a devouring force threatening the human order. Existential fears, particularly those originating in the seventeenth and eighteenth centuries, were projected onto the figure of 'the wolf' (for detailed information regarding Lusatia, cf. Hartstock 2017). Co-remembered echoes of these fears resurface today with each new pack dwelling in Eastern and Northern Germany. However, current discourse on wolves can certainly neither be reduced to scattered fears nor to a lack of respect concerning encounters with the predators. Instead, there are plenty of diverse narrations about 'the wolf' and wolves – as figures, species and individuals – concerning their return to Lusatia and their life in landscapes called 'cultured'.

Therefore, we focus on discursive agglomerations unfolding at the margins of traditional and contemporary narratives on the wolf and wolves. With the concept of *lieux de mémoire*, Nora developed a research approach that works as an analytical tool to deal with culture(s) of remembrance, viz. with an organized realization of the past via dunning or glorifying commemoration as it can be seen in memorials and their embeddedness in practice. The transfer of certain meanings and motifs onto either real or imaginary events, figures and places when researching narrative cultures is called "crystallization" (Köhler-Zülch 1996). In this sense, our lithic wolf on top of the hunting memorial at Laußnitz Heath emblematizes our multidimensional object of research: the wolf in its crystallized shape as a *lieu de mémoire*.

Threads and knots in the web of discourse

For us as researchers of narrative cultures, the Laußnitz wolf pillar, its temporal black dress and its use in local media mirrors several narratives that tell us about humans and wolves living in each other's neighbourhoods. In its original intention, it was meant as a remembrance of an exceptional hunting event unusual in its day. A wolf was shot there on 11 November 1740 for the first time in 56 years. It was an extraordinarily large specimen weighing 82 pounds.[9] Following the local legend recorded, "Electoral Prince August the III [sic] was attacked by a large wolf here in 1740. The ruler when confronted with the animal was said to be at great risk of death. Just at the right moment, a confident hunter [...] subdued the grim wolf with a perfectly aimed shot"[10] (Störzner 1904, 362). The inscription on the pillar does not mention this saga. It can be assumed that the story was only composed retrospectively: the ruler in a perilous situation and the dangerousness of the beast potentiate the memorability of place and event. Memorial and legend kept alive the memory of wolves dwelling in Laußnitz Heath 'once upon a time' during the nineteenth and twentieth century (Störzner 1904; Bies 2014). The fact that the remembered incident was an isolated case got lost in the collective memory – an occurrence significant for the selections and oversubscriptions inherent to the formation of a collective memory.

Despite this quite clear inscription labelling this wolf as a "first one", the pillar is a memorial for the last wolf shot there in the collective memory. Children and adolescents were asked to answer the question on the website *Nature Detectives*, hosted by the German Federal Office for Nature Conservation, whether they like it or find it rather uncanny that wolves are back in Germany. One of the responses published is given by 10-year-old Abigail:

> I find it great that there are more wolves in Germany again! If my parents and I go to visit my grandmother in Hoyerswerda, we pass by the wolf pillar at Laußnitzer Heide. My mum explained to me that this memorial is supposed to remind us of a hunt during which the last wolf in Saxony was killed. That makes me really sad and angry and I would find it cool if, despite that, we'd meet a free, living wolf in the forest.[11]

The discussion about first and last wolves goes on, as does the question regarding how long Germany was a country without wolves. Varying allegations with a vast range of wolfish absence from 50 to 200 years do not only express historical uncertainty, local difference or animal agency. Within the different narratives inherent in these numbers, the return of wolves is (de) naturalized, the focus shifts from a continuous migration of animals towards a rather stable and established system where wolves, due to their long absence, have lost their claim. These narratives, as contrary as they seem, have in common that they help to put the world in a specific order, each of them makes sense to their narrators (Bendix 1996).

A lithic wolf and mortal sheep

The question of return is also a quest for representation. Regarding the stone wolf in 1740 or taxidermists' wolves nowadays, respectively (cf. Arnold and Frank in this volume), the process of iconography is always a negotiation of prospective memory. Which animal is shown when people, for example, visitors to museums, hikers in landscapes shared again with wolves or readers of newspapers, ask for the image of a wolf? What images do we want future generations to see? The animal on top of the pillar is sitting, almost like a trained dog, the hair surrounding its face with the slightly opened snout is long and reminds one of a mane. Its tail is bushy and bent double upwards over the back. The body hair is short, invisible, and its legs are depicted with impressive muscles. What is this animal doing? It does not look like a defeated beast, or like a fighting or wounded wolf. The way it has sat there for almost 300 years makes it resemble more a guardian of the woods. If it came to life tomorrow, it would probably be categorized as a hybrid and shot immediately.

Now, however, since wolfish returnees dwell in Lusatia, the lithic wolf might be read as a family ancestor of the living wolves. When the statue wore a black hat and scarf in January 2018, journalists of a Saxon newspaper assumed that these must be sables. To imagine a wolf wearing sables means to anticipate a mourning wolf. We cannot tell if that indeed was the intention, but for us the narration is the fact of interest here. For the journalists, the wolf initially placed there to remind passers-by of a hunting event in 1740 is now perceived as a fellow of living wolves, or at least as an adequate context to communicate grief or concern related to the local wolf population. One year previously, in August 2017, the wolf was dressed in a fluorescent orange safety vest. When Stephan Kaasche, one of the referees of the contact office "Wölfe in Sachsen", who documented this scenario, uploaded a picture onto a social network, somebody commented, "A bulletproof one would be better".[12]

Thinking of grief[13] and concern as possible empathic utterances towards a wolf population is connected to narrations that seem odd considering the repertoire proliferated and conserved within collective memories. Following the outlines of 'the wolf' as a *lieu de mémoire*, grief and concern were rather thought of as an effect of wolfish action than as a feeling towards or even with wolves. When the Laußnitz pillar was erected in the 1740s, empathy with wolves would have been beyond (as opposed to within) the truth (Foucault 1981, 60).

Figure 3.2 Wolf in a safety vest
Photographer: Stephan Kaasche

While old static places such as the Laußnitz wolf pillar are altered in their readings and become plurivalent, there is also a need for new places that reassure a clear interpretation of 'the wolf'. Despite the fact that it is a hunting memorial, the Laußnitz pillar paradoxically shows the wolf instead of its pursuer and puts the defeated victim of a hunt on a lithic throne. The human hunter who shot the wolf remains invisible. Acknowledging the concerns of livestock owners and the so far unfulfilled calls for a 'regulation', i.e. minimization via shooting, of the local wolf population uttered by some of them, the pillar might seem provoking in a metaphorical way. The wolf sits high above, untouchable and sacrosanct, protected by federal and European law. The sculpture moves as much as the debates about a reduction of packs: not an inch. When victims of the grey returnees become temporary memorials, it can be read as a counter-motion. This happened in Großdöbschütz, Saxony, in April 2017. After he had heard a noise outside, an 18-year-old man found his lamb dead and its mother dying as the result of a wolf attack. Together with his grandfather, the owner placed the corpses in a wheelbarrow and put them on the sidewalk. The men used white chalk to write a message on the wooden plank of a fence behind the dead animals: "These sheep were killed by wolves."[14] The intention, as described by a journalist, was to create an alarming memorial.[15] Here the longing for an interpretative locking-up becomes obvious. The organic memorial can be read as a temporal attempt to (re)configurate space, meaning and memory of what and who wolves are. At the same time, new wolf-stones are erected, engraved with the date of return (as in Rochau Heath, Brandenburg) or first successful reproduction of wolves (in Daubitz lake district). Conflicting with the temporal memorial at Großdöbschütz, these stones memorize and narrate the return of wolves as a positively memorable event.

Johannes Fried, a historian and medievalist, conducted intense research on the value and reliability of memory performance as a source for historic investigations. He points out the reciprocity of memory and perception for the constitution of the cultural memory of communities. Following his argument, memory effects how and what is perceived by people and vice versa, each reformation of perception changes our memory ability. It is in this sense that Fried claims an "indivisible interplay of event, sensation, circumstances, perception, memory and the factual reproduction of the remembered as well as of the constellation retrieving the memory"[16] (Fried 2012, 107). Interests, emotions and prejudices channel the ability to perceive and remember as much as mental dispositions. Therefore, memories as sources are labelled "blurred" by Fried. He uses the metaphor of the *Veil of memory*, as he entitled his publication on "anthropological problems when considering the past".[17]

Figure 3.3 Year 2000 – Wolf-stone commemorating the first documented reproduction of wolves in Germany within the return process at Daubitz lake district
Photographer: Marlis Heyer

The selective effects of remembering always include forgetting. Forgetting can be described as a technique consistent with conscious or unconscious shielding, masking, losing, covering, ignoring, denying, overwriting or hiding; as well as of the neutralization of valences, disturbance, or confusion – as opposed to neuronal malfunction (Assmann 2016, 21–26). In other words, to forget is a creative process itself. It relieves our memory of the dead-weight of countless perceptions and impressions and, therefore, enables us to abstract experiences. Fried writes about a "subtle, culture-constituting oblivion"[18] (Fried 2012, 114). But what is going to be forgotten and what will be remembered? Getting back to our field of interest, we ask: How much and what kind of cultural knowledge about human–wolf relations are lost due to the lack of applicability during wolfish absence? And how much and which part of routines in human–wolf encounters have been covered by singular traumatic interspecies experiences and the heavy support and proclamation of the extirpation of wolves in the past?

Forgetting helps to simplify reality and, therefore, makes it easier to set up global narratives, truths and dichotomies. The phenomenon to simplify narratives in retrospective and to declare that things were clearer in former times and understandings of the world can be observed concerning many topics; the construction of a past ignores the original complexity of a former present. Along with nostalgia, narrations about past knowledge imply that our cultural ancestors handled things in a more 'natural', easier and better way. These metanarratives miss the fact that traditional narrations also bear an incredible amount of intricacy. The simplification is a process that happens afterwards, filtering out puzzling or disturbing aspects and, thus, shaping our collective memory. Therefore, the question of how to deal with wolves today is also a question of how we remember what our ancestors did during the past.[19] The past we want to remember collectively, as such, also shapes our multispecies future(s).

A glimpse at the veil of memory

To come full circle, we want to refer once again to the articles with which we opened our paper: these kinds of short reports are typical news on behalf of the Lusatian wolf population. Editors receive press releases from local institutions and authorities and add some content with regional specification and editorial comments. Altogether, recipients of the *Sächsische Zeitung* (SZ BZ 2017a, 2017b, 2017c, 2017d, 2017e) are well-informed about how their wolfish neighbours are thriving. Reading a local newspaper in Lusatia means to keep on the wolves' track. Since the return of the wolves, the media has constantly covered both their visible or traceable activities and the emotions they induce.[20] Short reports similar to the ones with which we opened our paper are combined with extensive news, reports, leading articles and cartoons. The contributions represent the voices of different stakeholders, such as nature conservationists, pro-wolf campaigners, politicians, employees of Saxon state forests, hunters, livestock owners, the contact office "Wolves in Saxony", political parties and the citizen initiative *"Wolfsgeschädigte und Besorgte Bürger"*[21] founded in 2017. Clearly these voices intermingle and intersect and are often conflicting. Headlines such as "Cull or preserve"[22] (SZ BZ 2017d, 7) invoke direct positions among readers (and writers), echoing in columns and letters to the editor. Altogether, the frequent attendance of narrated wolves in local media shows the high popularity of the topic, involving repeating patterns and specific imagery. Many of the announcements prefer to talk about 'the wolf' as a seemingly well-known prototype of a species whose 'nature' is sufficiently explored. References to this foundational knowledge are historic sources such as the encyclopaedia *Illustriertes Thierleben. Eine allgemeine Kunde des Thierreichs* [Illustrated Animal Life. A General Account of the Animal Kingdom] (Brehm 1864–1869) and collectively memorized legends and stories transporting discourses and narratives and densifying in numerous motives and variations of the topos of the 'big bad wolf' (or is it just 'the wolf'?).

Preoccupation with memory and oblivion leads to different tasks concerning research on cultural history. One of them would be to 'raise the veil' and separate facts from fiction, originals from forgeries. Opposed to this approach, we want to focus on the veil itself, its fabrication and texture. What is it made of? Which patterns are woven into the tissue and which knots and entanglements help to keep it resilient? The perspective of understanding history as "consciously remembered past"[23] (Fried 2012, 53) which is constantly morphing under the pressure of the present, as Fried puts it, fits well with Nora's approach, who coined the term of a historiography "of second degree". This historiography, instead of focusing on the events themselves, asks for semantic changes and shifts of these events and their commemoration throughout époques. Thus, Nora's historiography is not about "the past, as it really was, but about its permanent recycling, its use and abuse and its meaning for consecutive presents; not about traditions, but about the ways of producing and imparting them"[24] (Nora 2005, 16; as quoted in Siebeck 2017, 4 f.).

How to narrate wolves?

The return of wolves to Germany is embedded in an accompanying setting called 'wolf management'. This setting goes beyond monitoring, registering and analysing wolves and their behaviour(s).[25] It also tackles some of our traditions of narrating wolves, or rather 'the wolf': through re-narration, wolves are implemented within a seemingly factual, modern, scientific and unemotional discourse. Their former place within narrations in fictional and overtly openly emotional discourses closely connected to contexts of rather traditional narratives and collective memory is labelled as inadequate.

The strong aim to resettle narratives about wolves into a clear, authorized discourse is, for example, mirrored by the discussion whether the wolf belongs in the game law or not. Saxony is the only federal state in Germany which included wolves in the game law in 2012. Despite a year-long closed season for the wolf, its implementation into another bureaucratic and institutionalized framework has effects on how to handle local wolf populations. European and the National Nature Conservation Act must be consulted for interactions outside the normal, as well as the federal game law and the Saxon plan for wolf management. Wolves and human–wolf interactions narrated in these contexts – as different as they might be – demystify wolves; they replace emotions with regulations, questions with instructions. These procedures of replacement do not absorb affective approaches towards wolves. Emotions such as fascination or fear do not disappear, but they lose their legitimate place in authorized debates about wolves.

Terms such as 'fairy tale' work as an accusation towards the contrary party in conflicts concerning human–wolf coexistence in Germany: while advocates of the return of wolves blame sceptics for 'still believing in the tale of Little Red Riding Hood', respectively suffering from 'Little Red Riding Hood syndrome',[26] their opponents accuse their counterparts of trusting in 'the tale of the shy wolf'[27]. The term 'fairy tale' helps the disputants to discredit each other, taunting naivety, infantile behaviour and worldviews. 'To believe in fairy tales' equals talking about wolves in the wrong way – or is it the assumed plainness of the wolf's role in traditional narratives that makes each side uncomfortable to settle themselves in a line with 'classic' wolf narratives? As shown in other contributions to this volume, the simple idea of a pro/ contra set in the debates aroused with and around wolves is far too simple and does not explain sufficiently how wolves came to take such a crucial role in negotiation processes. Is this dichotomous way of conceptualizing our relationship with wolves a heritage of traditional narrating?

It seems to be true: following representations of wolves and the human–wolf relations in mass media, there seem to be only two reliable collectively remembered sources to explain our difficult connection: *Little Red Riding Hood* and (far rarer) *The Wolf and the Seven Young Goats*. Despite this popularity ranking, the picturing of wolves has never been one-dimensional nor simply simplifying. Instead, traditional narratives of wolves offer a complex and complicated portrait of wolves and wolfish characters. Society's need for easy explanations, allowing one to ignore complexity, robs us of a rich cultural heritage of wolf narrations for the sake of a clear approach and readability.

While acknowledging that the cultural process of simplification is crucial, since otherwise, no such thing as a cultural identity or collective memory would be able to exist (as described above), we are curious about if and how we can bring some of the 'lost' and/or forgotten complexities back into play, letting them help us to rethink and reimagine wolves and our coexistence. The contact office "Wolves in Saxony" at Rietschen decided to include different narrative approaches towards wolves in their permanent exhibition about the new neighbours. A big wooden book can be found which includes fables and folk tales as well as extracts from classics such as *The Jungle Book* by Rudyard Kipling (1894) in the garden of the exhibition. Thus, wolves are shown as embedded into different social, cultural and historical contexts all over the globe. Reading these short stories, wolves appear and are narrated as familiar and strange, wild and tame, dangerous and helpful.

If we approach the wolf as a *lieu de mémoire*, it is obvious that this crystallized figure is one in motion. It has cracks, is deformed and reshaped by time, social and cultural developments and processes. As a *lieu de mémoire*, the wolf is ambivalent and ambiguous, it combines contesting narratives and interpretations of the place of wolves and humans in a shared space, of relations between imagined wilderness and proclaimed cultural heritage, of tensions of practices, humanly and other-than-human. Following current debates about wolves' space in Germany, one might even ask if we still can assume 'the wolf' as one single collective memory or if we are talking about multiple *lieux de mémoire*, separated distinctly from each other. Here, the concept of palimpsest (cf. Kirschbaum 2016) might help us to frame the contested *lieu de mémoire*. Analysing the collective memory of, on and about wolves, 'the wolf' today means to analyse a fight for interpretative control. The wolf is a strongly contested space within our society's capacity to remember, narrate and interpret – itself, its inclusions and exclusions and its relationships to others.

Thinking about Nora's theoretical approaches towards collective memory, one might wonder if the wolf, contested as it is nowadays due to its 'comeback' in both literal and metaphorical ways, can really be considered as a *lieu de mémoire*. As outlined above, discussions on wolves are vivid, and the question who and what we are talking about while talking about wolves is not easy to answer. When Nora stated that a *lieu de mémoire* replaces a *milieu de mémoire*, we wonder if this procedure can be inverted. According to our observations, the return of wolves to Germany stimulates the discourse, alters imaginations and questions implicitness. Thus, a former crystallized figure becomes cracked and reshaped. The non-human actors themselves reanimate the narrative culture(s) of the spaces into which they return.

Thus, 'the wolf' as a *lieu de mémoire* must be conceptualized as a bundle of practices, characterized by its instability and permeability instead of stability. The return of wolves is not just a return into and renewal of contemporary medial representations; it is as much a return into everyday lives. One possible approach towards new actors is monitoring. Embedded in a socio- and eco-biopolitical framework called 'wolf management', the behaviour of wolves is documented and administrated. Wolf management, as a broad project, tries to meet the concerns of both poles: people and wolves. Therefore, 'to manage', on the one hand, can be understood as 'to care': this meaning narrates wolves as a species deserving protection and our society as one that should aim towards a sustained, strong and assured wolf population. Following this narrative trace, a repeated decline of the wolf population is something that wolf management wants to avoid. On the other hand, the necessity to manage wolves narrated within the term is an implicit acknowledgement of (potential) danger(ousness) and arouses the image of wolf management as an instrument to protect people from wolves and curtail the uncontrolled aspects of wandering, breeding and feeding of the non-human migrants. The term management also reinforces the idea that it is humans who control their surroundings, may those be of a 'cultural' or 'natural' character. And, despite the implications of the term 'wolf management' or, in a broader perspective, 'wildlife management', the concepts and plans address human actors and institutions; as much as it is 'wolf management', it is also 'human management'.

The return of wolves into Lusatian landscapes, from heathlands and military grounds to lake districts to former pit mines, is reported with the greatest interest not only based on sceptical curiosity but also in another narrative: where there are wolves, there must be 'intact nature'. This topos, structured by an underlying dichotomy of nature and culture inherent to Western modernity (cf. Latour 1993), reflects diffuse ideas of a seemingly closed, self-sufficient space that exists apart from human influence. This space, in some contexts even connoted with paradise, is often narrated as needing and deserving protection. Surely this vision of 'nature' designs a space of longing and desire; the more contested, the more threatened it seems to be (cf. Uekötter 2014; especially for the 'German forest' as memory space: Lehmann 1999, 2001, 2010). While wolves and their dwelling in Lusatia are associated with a positive, even idyllic imagery of 'nature', on the one hand, there is the topos of wolves bearing 'wilderness', on the other hand. 'Wilderness' is often narrated as being contrary to dystopian scenarios of the Anthropocene in current 'rewilding' discourse and is loaded with positive, reviving connotations. Nevertheless, the story might be different regarding Lusatia. As an area located at the periphery from many angles (socio-geographical, demographical, economical, just to name a few), wilderness sounds rather like wasteland. While some journalists proclaim that wolves, along with the lynx, beaver and eagle, transform wastelands into landscapes of hope (Seewald 2012), rather more bitter voices are not far away: "Here, in the vast wastelands between Cottbus and Zittau, he [the wolf] dared to venture for the very first time; in the beginning shy and fearful, but soon with more courage [...] Unfortunately, in Lusatia he only finds what he also had to make do with in fairy tales: grandmothers"[28] (Bittrich 2016, 180).

Conveyed and current narratives about the wild animal in our neighbourhood

While the animals in their bodily form were extinct, the *lieu de mémoire* remained; so did the memorials of their disappearance, such as the Laußnitz wolf pillar, but not without being altered. Therefore, the animal coming back today does not only find itself in a landscape different to the one where its species was extirpated, but the animal itself is also a different one (Tschofen 2017, 8). During their physical absence, some of the narratives shaping our understanding of wolves were filled with new facets, different interpretations and possible understandings of these large carnivores. While acknowledging the age of the Anthropocene, the wolf mutated somewhat into a *lieu de promesse* for parts of German society. Narrated as an essence of wilderness and, therefore, as an image of an intact (e.g. humanless) nature, the first wolves were welcomed and recited as a "miracle" (Fuhr 2014, 11). As a literal "return of the wild", respectively "wild returnees", the first German wolf packs were appreciated and monitored at once, as can be seen in the media coverage and scientific reports (cf. Heitkamp 2004). The idea(l) of wilderness as contrary to agriculturally shaped cultural landscapes and the massive human impact on ecosystems changed the narratives of extinction: triumph became defeat and the mastering of nature turned into loss. Under these circumstances, the Laußnitz wolf pillar, a memorial of nature's defeat by humans and, therefore, an icon for the human manageability of nature, could be transformed into a place significant for a deficit. Related to his wolfish relatives, dwelling and alive in the surrounding area, the lithic wolf becomes a mediator between human and wolfish inhabitants and their interests. His temporary garments, such as sables or a safety vest, express conflicts about the contested space of wolves in our collective memories and understandings.

However, in some situations, it seems as if not only human actors struggle with the overwhelming supply of different possible approaches towards wolves, but the wolves also seem to be trapped in or even lost in interpretation. How can they be wild and controllable at the same time; how could they hide inside forests, while their prey grazes on meadows and feasts in fields; how can they be afraid of human beings, while they meet them in nearly every corner of their possible habitats, use their infrastructure and share spaces with them? Even complaints about wolves not being afraid of the sound of rifles during a hunt in their territory seem odd when reflecting that many wolves in Germany find refuge at military training grounds, some of them hosting more than 200 training sessions each year.

Our society is one that struggles constantly with ambiguity (Bauer 2011). The culturally inscribed fetish for clarifications and classifications makes it hard to accept ambiguous spaces within the collective memory – despite the fact that this might conjure "hybrids, monsters", as described by Latour (1993, 47). With wolfish actors present, a single interpretation and a clear position towards these animals has become even more impossible than it is by definition: Despite complex interplays between wolves and humans in the past, 'the wolf' could crystallize as a – more or less – defined *lieu de mémoire*, preconditioned by collective dynamics of remembering and forgetting. The absence of wolves guaranteed, at least to a certain degree, a stability of their image. Nowadays, not only does the wolves' presence challenge our society's capacity to coexist with animals that were extinct previously. Now that international and national law protects them, it also brings wolfish agency into play (as described by Arnold in this volume) and depredates humans' narrations of their monopoly. It is far more complicated to narrate collectively and, therefore, remember something present, alive and agile – wolfish actions have to be included into the complex project of (re-)narrating wolves and our relation towards each other.

Notes

1 Local editions for Kamenz (SZ KM 2018) and Großenhain (SZ GRH 2018).
2 All quotations from sources that were only available in the German language were translated by the authors. Originals can be found in the endnotes. "Das Tier aus Stein trägt zurzeit Schwarz."
3 "Kein Grund zum Heulen."
4 "Dokumentations- und Beratungsstelle des Bundes zum Thema Wolf", founded in 2016, is the institution responsible for collecting and documenting data of all federal wolf management and serves as an information centre for wolf-related topics.
5 The statistical surveys for the wolf monitoring year 2017 were not concluded when the paper was written.
6 Lusatia (German: Lausitz, Sorbian Łužyca/Łužica) is a region in Central Europe, traversed by the rivers Spree and Neisse. Lusatia stretches from southern Brandenburg to eastern Saxony and the Polish voivodships of Lower Silesia and Lubusz.
7 The office was founded in 2004 by the Saxonian Ministry of the Environment and Agriculture and the district of Lower Silesian Upper Lusatia (Ministerium für Umwelt, Gesundheit und Verbraucherschutz des Landes Brandenburg 2011, 65).
8 "Am wilden Wolf konnten sich, wie bei den Märchenwölfen der Brüder Grimm, alle Folter- und Totschlaggelüste der Menschen austoben."
9 A similar monument can be found between Moritzburg and Weinböhla: it commemorates a battue attended by Electoral Prince Johann Georg I that took place in 1618. The sandstone sculpture was created by Dresdner sculptor Sebastian Walther (1576–1645). The fact that it was restored under the reign of Johann Georg II in 1672 marks that these kinds of pillars were celebrated and cherished as memory spaces during the seventeenth century (Gurlitt 1904, 123–124).
10 "hier der Kurfürst August III. [sic] im Jahre 1740 von einem großen Wolfe angefallen worden sein [soll]. Der Landesfürst sei dadurch in die größte Lebensgefahr gekommen. Noch im rechten Augenblick habe ein sicherer Schütze [...] den grimmen Wolf durch einen wohlgezielten Schuß erlegt." The ruler mentioned, August III, reigned as Friedrich August II as long as he was electoral prince. His title changed to August III when he became King of Poland in 1734.
11 "Ich finde es super, dass es in Deutschland wieder mehr Wölfe gibt! Wenn ich mit meinen Eltern zu meiner Oma nach Hoyerswerda fahre, kommen wir immer an der Wolfssäule in der Laußnitzer Heide vorbei. Meine Mama hat mir erklärt, dass uns das Denkmal an eine Jagd erinnern soll, bei der der letzte Wolf Sachsens getötet wurde. Das macht mich echt traurig und wütend und ich fände es cool, wenn wir jetzt trotzdem irgendwann einem frei lebenden Wolf im Wald begegnen würden." NaturDetektive (2019).
12 "Eine kugelsichere wäre besser."
13 Cf. Cunsolo and Ellis (2018) for a conceptualization of ecological grief.
14 "Die Schafe wurden von Wölfen getötet."
15 https://www.focus.de/regional/videos/wie-konnte-das-passieren-wolf-reisst-scha fe-in-sachsen-besitzer-stellt-tote-kadaver-auf-schubkarre-aus_id_6983701.html
16 "untrennbare[n] Wechselwirkung zwischen Ereignis, Sinneseindruck, Rahmenbedingungen, Wahrnehmung, Erinnerung und sachlicher Reproduktion des Erinnerten, aber auch der die Erinnerung abrufenden Konstellation."
17 Subtitle of the Annual Lecture at the German Historical Institute in London held by Fried in 1998. https://www.ghil.ac.uk/publications/annual_lectures.html [accessed 4 June 2018].
18 "schleichendes, kulturstiftendes Vergessen."
19 In February 2018, the Altmark-Zeitung published a comment of one of their readers about the discussion on wolves titled "Our ancestors weren't stupid" [Unsere Vorfahren waren doch nicht dumm] (AZ 2018).

20 While Google Analytics shows a stable distribution of news concerning wolves during the past few years, our evaluation of the printed version of *Sächsische Zeitung* (local edition for Bautzen) shows an increase since 2015.

21 The initiative's name is hardly translatable; it might be paraphrased as "Wolf-damaged and concerned citizens". It is noticeable that the term *"Besorgte Bürger"* was also used in an inflationary manner as a (self-) description by the right-wing protests which have been taking place in many German cities since 2014 with Dresden as a starting point. The authors want to follow traces of parallel narratives in discourses on wolves and migration in further investigations.

22 "Schießen oder schützen?"

23 "bewusst erinnerte Vergangenheit."

24 "die Vergangenheit, wie sie eigentlich gewesen ist, sondern [um] ihre ständige Wiederverwendung, ihren Gebrauch und Missbrauch sowie ihren Bedeutungsge-halt für die aufeinanderfolgenden Gegenwarten; nicht [um] die Tradition, sondern die Art und Weise, wie diese geschaffen und weitergegeben wird."

25 A broad and open conceptualization and understanding of wolf-management, for example, as a multispecies co-learning process, is a fundamental part of the research design of the project "Die Rückkehr der Wölfe. Kulturanthropologische Studien zum Prozess des Wolfsmanagements in der Bundesrepublik Deutschland" [The return of wolves. Cultural anthropological studies concerning the process of wolf-management in the Federal Republic of Germany], led by Prof. Dr. Michaela Fenske, funded by the Deutsche Forschungsgemeinschaft (DFG). One of the authors, Marlis Heyer, conducts her PhD project as one of two sub-projects in this setting.

26 Cf., for example, an official statement of the contact office "Wolves in Saxony" from 2007 titled "Rotkäppchensyndrom feiert Auferstehung", a dementing hoax concerning wolves, or an article written by Steffen Winter (2008), titled "Das Rotkäppchen-Syndrom", reporting on the current wolf–human situation in the Lusatia region.

27 As a phrase, 'the tale of the shy wolf' pops up in debates and public negotiations, for example; in an article published in the WELT newspaper column "Wissen": "Das Märchen vom lieben Wolf" (Fuhr 2012) or reader's comment "Das Märchen von dem scheuen Wolf" (de Haan 2017).

28 "Hier, in der weiten Ödnis zwischen Cottbus und Zittau, hat er [der Wolf] sich zu allererst ans Licht gewagt, anfangs noch scheu und furchtsam, bald mutiger, als ihm signalisiert wurde, er sei herzlich willkommen und könne sich auf ein Rundum-Sorglos-Paket freuen. [...] Leider findet er in der Lausitz typischerweise nur das vor, womit er auch im Märchen vorlieb nehmen musste: Großmütter."

Bibliography

Assmann, A. 2016. *Formen des Vergessens*. Göttingen: Wallstein.

*AZ*2018. "Comment regarding 'Unsere vorfahren waren doch nicht dumm'." *Altmark-Zeitung*. https://www.az-online.de/leserbriefe/altmark/unsere-vorfahren-waren-doch-ni cht-dumm-9625559.html [accessed 29 March 2018].

Bauer, T. 2011. "Kulturelle Ambiguität." In *Die Kultur der Ambiguität. Eine andere Geschichte des Islams*, edited by T. Bauer, 26–53. Berlin: Verlag der Weltreligionen.

Bendix, R. 1996. "Zwischen Chaos und Kultur. Zur Ethnographie des Erzählens im ausgehenden 20. Jahrhundert." *Zeitschrift für Volkskunde* 92 (2), 169–184.

Bies, W. 2014. "Wolf." In *Enzyklopädie des Märchens. Handwörterbuch zur histor-ischen und vergleichenden Erzählforschung*, Vol. 14, edited by R. W. Brednich, H. Alzheimer, H. Bausinger, W. Brückner, D. Drascek, H. Gerndt, I. Köhler-Zülch, K. Roth, and H.-J. Uther, 912–923. Berlin: De Gruyter.

Bittrich, D. 2016. *99 deutsche Orte, die man knicken kann.* Reinbek bei Hamburg: Rohwolt.

Brehm, A. 1864–1869. *Illustriertes Thierleben. Eine allgemeine Kunde des Thierreichs,* 6 vol. Hildburghausen: Verlag des Bibliographischen Institutes.

Cunsolo, A. and N. Ellis. 2018. "Hope and Mourning in the Anthropocene: Understanding Ecological Grief." https://theconversation.com/hope-and-mourning-in-the-anthropocene-understanding-ecological-grief-88630 [accessed 20 April 2018].

de Haan, H. 2017. "Das Märchen von dem scheuen Wolf." https://www.az-online.de/leserbriefe/isenhagener-land/maerchen-scheuen-wolf-8234604.html [accessed 20 April 2018].

den Boer, P., H. Duchhardt, G. Kreis, and W. Schmal, eds. 2012. *Europäische Erinnerungsorte,* 3 vol. München: Oldenbourg Verlag.

Foucault, M. 1981. "The Order of Discourse." In *Untying the Text: A Post-Structuralist Reader,* edited by R. Young, 48–78. London: Routledge.

François, É. and H. Schulze, eds. 2001. *Deutsche Erinnerungsorte,* 3 vol. München: C. H. Beck.

Fried, J. 2012. *Der Schleier der Erinnerung. Grundzüge einer historischen Memorik.* München: C. H. Beck.

Fuhr, E. 2012. "Das Märchen vom lieben Wolf." https://www.welt.de/print/die_welt/wissen/article111959349/Das-Maerchen-vom-lieben-Wolf.html [accessed 20 April 2018].

Fuhr, E. 2014. ²2015. *Rückkehr der Wölfe. Wie ein Heimkehrer unser Leben verändert.* München: Riemann Verlag.

Gurlitt, C. 1904. *Beschreibende Darstellung der älteren Bau- und Kunstdenkmäler des Königreichs Sachsen. Heft 26, Amtshauptmannschaft Dresden-Neustadt (Land).* Dresden: C. C. Meinhold & Söhne.

Hartstock, E. 2017. *Von Feuerbrünsten, Epidemien und Dürrekatastrophen. Die Geschichte der Plagen in der Oberlausitz. Eine Chronik der Ereignisse von 1112 bis 1869.* Spitzkunnersdorf: Oberlausitzer Verlag.

Heitkamp, S. 2004. "Lausitzer Wölfe auf dem Rückzug." https://www.lr-online.de/nachrichten/lausitzer-woelfe-auf-dem-rueckzug_aid-4143232 [accessed 20 April 2018].

Hiltmann, H. 2013. "'Guter Wolf – Böser Wolf'. Die Ambivalenz norröner Wolfsbilder und ihrer persönlichkeitsstiftenden Funktion im europäischen Vergleich." In *Anima in fabula. Interdisziplinäre Gedanken über das Tier in der Sprache, Literatur und Kultur,* edited by M. Ulrich and D. De Rentiis, 143–174. Bamberg: University of Bamberg Press.

Holbek, B. 1984. "Epische Gesetze." In *Enzyklopädie des Märchens. Handwörterbuch zur historischen und vergleichenden Erzählforschung,* vol. 4, edited by R. W. Brednich, H. Bausinger, W. Brückner, L. Röhrich, K. Roth, R. Schenda, I. Köhler, U. Marzolph, E. Moser-Rath, and K. Ranke, 58–69. Berlin: De Gruyter.

Kipling, R. 1894. *The Jungle Book.* London: Macmillan.

Kirschbaum, H. 2016. *Im intertextuellen Schlangennest. Adam Mickiewicz und polnisch-russisches (anti-)imperiales Schreiben.* Frankfurt a.M.: Peter Lang.

Köhler-Zülch, I. 1996. "Kristallisationsgestalt." In *Enzyklopädie des Märchens. Handwörterbuch zur historischen und vergleichenden Erzählforschung,* vol. 8, edited by R. W. Brednich, H. Bausinger, W. Brückner, L. Röhrich, K. Roth, R. Schenda, I. Köhler, U. Marzolph, E. Moser-Rath, and K. Ranke, 460–466. Berlin: De Gruyter.

Kreis, G. 2010. "Pierre Nora besser verstehen – und kritisieren." In *Schweizer Erinnerungsorte. Aus dem Speicher der Swissness,* 327–347. Zürich: Verlag Neue Zürcher Zeitung.

Latour, B. 1993. *We Have Never Been Modern*. New York, NY: Harvester Wheatsheaf.

Lehmann, A. 1999. *Von Menschen und Bäumen. Die Deutschen und ihr Wald*. Reinbek bei Hamburg: Rowohlt.

Lehmann, A. 2001. "Der deutsche Wald." In *Deutsche Erinnerungsorte*, vol. 3, edited by É. François and H. Schulze, 187–200. München: C. H. Beck.

Lehmann, A. 2010. "Der deutsche Wald." In *Waldeigentum. Dimensionen und Perspektiven*, edited by O. Depenheuer and B. Möhring, 3–19. Heidelberg: Springer.

Ministerium für Umwelt, Gesundheit und Verbraucherschutz des Landes Brandenburg, ed. 2011. "Mit Wölfen leben. Informationen für Jäger, Förster und Tierhalter in Sachsen und Brandenburg." Potsdam: Referat Umweltinformation, Öffentlichkeitsarbeit.

NaturDetektive 2019. "Eure Meinung zählt! Abigail (10) aus Dresden." Bundesamt für Naturschutz. https://naturdetektive.bfn.de/lexikon/tiere/saeugetiere/woelfe-wolf/eure-meinung-zaehlt.html

Nora, P. 1984–1994. *Les lieux de mémoire*, 7 vol. Paris: Gallimard.

Nora, P. 2005. "Wie lässt sich heute eine Geschichte Frankreichs schreiben?" In *Erinnerungsorte Frankreichs*, edited by P. Nora, 15–23. München: C. H. Beck.

Nora, P. and L. D. Kritzman, eds. 1996. *Realms of Memory: Rethinking the French Past*. New York, NY: Columbia University Press.

Olrik, A. 1909. "Epische Gesetze der Volksdichtung." *Zeitschrift für deutsches Altertum und deutsche Literatur* 51, 1–12.

Parry, M. 1971. *The Making of Homeric Verse. The Collected Papers of Milman Parry*, edited by A. Parry. Oxford: Oxford University Press.

Rheinheimer, M. 1995. "Die Angst vor dem Wolf. Werwolfsglaube, Wolfssagen und Ausrottung der Wölfe in Schleswig-Holstein." *Fabula* 36 (1/2), 25–78.

Schenda, R. 1995. *Who's who der Tiere. Märchen, Mythen und Geschichten*. München: C. H. Beck.

Seewald, B. 2012. "Das absurde Lamento über die Wüstungen im Osten." https://www.welt.de/debatte/article109382875/Das-absurde-Lamento-ueber-die-Wuestungen-im-Osten.html [accessed 20 April 2018].

Siebeck, C. 2017. "Erinnerungsorte, Lieux de Mémoire, Version: 1.0." *Docupedia-Zeitgeschichte*, 2 March. http://docupedia.de/zg/Siebeck_erinnerungsorte_v1_de_2017. http://dx.doi.org/10.14765/zzf.dok.2.784.v1 [accessed 20 April 2018].

Störzner, F. B. 1904. "Die Laußnitzer Heide." In *Was die Heimat erzählt. Sagen, geschichtliche Bilder und denkwürdige Begebenheiten aus Sachsen*, 361–362. Leipzig: Arwed Strauch.

*SZ BZ*2017a. "Leserbriefe." *Sächsische Zeitung für Bautzen, das Oberland und das Heide-Teichland* (8 May), 15; (13 June), 18; (9 October), 15; (10 October), 8; (18 October), 15.

*SZ BZ*2017b. "Die, die den Wolf fürchten." *Sächsische Zeitung für Bautzen, das Oberland und das Heide-Teichland*, 24 January, 3.

*SZ BZ*2017c. "Wenn der Wolf am Wochenende zuschlägt." *Sächsische Zeitung für Bautzen, das Oberland und das Heide-Teichland*, 25/26 March, 24.

*SZ BZ*2017d. "Schießen oder schützen? Landrat Michael Harig plädiert für eine Regulierung des Bestandes. Jana Endel vom Wolfsbüro ist dagegen." *Sächsische Zeitung für Bautzen, das Oberland und das Heide-Teichland*, 22 April, 7.

*SZ BZ*2017e. "Was tun, wenn der Wolf kommt?" *Sächsische Zeitung für Bautzen, das Oberland und das Heide-Teichland*, 1 September, 5.

*SZ GRH*2018. "Steinerner Wolf trägt Trauer." *Sächsische Zeitung für Großenhain*, 23 January.http://www.sz-online.de/nachrichten/steinerner-wolf-traegt-trauer-3865658. html [accessed 20 April 2018].

*SZ KM*2018. "Laußnitzer Wolf trägt Trauer." *Sächsische Zeitung Kamenz*, 17 January. http://www.sz-online.de/nachrichten/laussnitzer-wolf-traegt-trauer-3861327.html [accessed 20 April 2018].

Tschofen, B. 2017. "Der Wolf ist da. Warum seine Wiederkehr ein Menschenthema ist." In: *Der Wolf ist da. Eine Menschenausstellung*, edited by Alpines Museum der Schweiz (B. Hächler) and Universität Zürich – ISEK (B. Tschofen), 5–10. Bern: Alpines Museum der Schweiz.

Uekötter, F., ed. 2014. *Ökologische Erinnerungsorte*. Göttingen: Vandenhoeck & Ruprecht.

Winter, S. 2008. "Das Rotkäppchen-Syndrom." http://www.spiegel.de/spiegel/a-585259. html [accessed 20 April 2018].

4 The story of *Wanderwolf*

A contested tale on the re-emergence of 'new wilderness' in the Netherlands

Anke Tonnaer

Introduction

The protagonist of this chapter is the lone wolf that crossed onto Dutch land in early March 2015 and subsequently drifted through villages, suburbs and along the Dutch–German border for a few days. Its life ended abruptly about a month later, when it was run over by a car back in Germany. The *Wanderwolf*, as the animal was soon baptized in the media, travelled back to the Netherlands at the beginning of 2016.[1] After firm lobbying to proclaim the creature as Dutch by the organization *Wolven in Nederland* (Wolves in the Netherlands)[2] – the *Wanderwolf*, thus, returned to the Netherlands in a freezer to its final destination at the Museum Naturalis in Leiden. It was then prepared and stuffed, so it could be exhibited to the public and possibly used for scientific research for the rest of its afterlife.

The *Wanderwolf* was the first wolf that could be identified with certainty as a wolf and, thus, announced the return of the species after the last one had vanished from the Dutch landscape about 150 years ago. Other earlier claims of wolf sightings were never confirmed, and the wolf that was found dead alongside a Dutch highway in 2013 had been planted there after it was shot in Eastern Europe.[3]

The *Wanderwolf* unleashed a national, highly mediatized outburst of responses. Indeed, as others have noted, the reappearance of wolves throughout Europe has led to conflicts, particularly in rural areas (Skogen et al. 2008). The heated reactions to the seemingly approaching wild in the Netherlands resembled the uproar wolves cause elsewhere. The *Wanderwolf* brought with it a value-laden narrative as an example of the species recovery which looks like stories others have noted (Drenthen 2015; Skogen et al. 2008). Studies conducted in, for instance, the French Alps, Germany, Finland and Italy reveal not only the existence of several successfully living packs of wolves but also the power politics between various parties and stakeholders, such as environmentalists, farmers and shepherds. Policy developments regarding this topic have recently been studied in a large European Research project on biodiversity (see www.biomot.eu).

However, the conditions of the *Wanderwolf*'s visit are, I suggest, introducing another chapter, i.e. a new modern tale, in the re-emergence of wilderness in Europe. Much of the Dutch media discussion was spun out on the double-edged sensation the *Wanderwolf* caused. An overwhelming enthusiasm about the apparent success of the new nature management, on the one hand, and an equally strong expression of fear and concern about the safety of humans (children in particular) and animals, both farmers' stock and pets, and especially for the wolf itself, on the other. How could wolves survive in a country the size of the Netherlands, which is run through by motorways and railway tracks and shared by about 17 million people? Indeed, the *Wanderwolf* also provoked a somewhat cynical response; sceptics asking whether the Netherlands had moved into a modern retelling of Little Red Riding Hood. In addition to embodying a promise of the recovery of the wild, it, thus, equally sparked a strong disbelief, because there is no 'real wilderness' in the Netherlands. From that perspective, the *Wanderwolf* embodied an anomaly, a fiery matter out of place (cf. Drenthen 2015).

Although the media storm quickly subsided following the disappearance of the *Wanderwolf*, the effects of its Dutch visit still linger. In this chapter, I study what this particular wolf has come to signify culturally in the contemporary public imagination and understanding of the Dutch landscape. I suggest that the *Wanderwolf* 'wandered' into the production of a new landscape ideology, i.e., another genesis of the Dutch landscape, in which it fulfils a different kind, yet equally controversial, iconic role compared to those we know from West-European ethnological and folkloric traditions (Van der Meulen 2016). The *Wanderwolf* heralded a modern fairy tale to the avid rewilder and nature lover. For others, the animal may have epitomized the announcement of the loss of ground, both literally and figuratively, within the borders of the nation that over the centuries had so successfully sculpted a disenchanted and demystified, expressively man-made landscape (Zwart 2003).

My reference here to the concept of a fairy tale is of course metaphorical. The *Wanderwolf* became the legend-like and equally ambiguous creature onto which different meanings could be canvassed. Socio-cultural perspectives on human–animal relations and human–nature relationships may be distilled more broadly in the way wolves have appeared in narratives and tales through time. Moreover, folk-tales reflect moral and ethical values as guidelines for society, with the wolf usually in the role of a significant adverse pendant. I will show that, building on what Drenthen (2015) has argued earlier, the present-day story of the wolf in the Netherlands epitomized the contemporary struggle of our place in the environment, in which the allocation of agency is not limited to humans and, consequently, upsets the idea of control in a country that we have become accustomed to seeing as an "essentially *humanized* environment" (Drenthen 2015, 322; italics in original; see also Tsing 2014).

I will start by sketching the political-environmental context in which the *Wanderwolf* was both enthusiastically welcomed and fiercely scorned. In particular, its presence may be best understood in the light of the larger eco-experiment that is occurring in one Dutch national park, the Oost-vaardersplassen, which, like the return of the wolf, is not without controversy, about which I have already written extensively elsewhere (Tonnaer 2014). Indeed, both the park and the wolf indicate particular readings of wilderness that have a long cultural history. Subsequently, I will discuss the opposing perspectives in more detail. I conclude with an analysis of the 'bewildering' dynamics of hosts and guests that I see emerging in the debate on the return of the wolf. In other words, the *Wanderwolf* points to unstable categories of who or what are the hosts and who or what are the visitors in the Dutch landscape.

In doing so, I aim to elaborate on three dimensions that can be discerned in the ambiguities of the human–nature nexus in which the *Wanderwolf* took centre stage. Firstly, the discourse surrounding the *Wanderwolf* alludes to a neoliberal logic through which the Dutch (but more broadly Western) environment seems to be increasingly shaped into a malleable, spectacular landscape. Secondly, related to the increasing commodification of nature, a widening discrepancy is emerging between experts and the lay public in terms of the authority of each to express and be heard in the debate about land management and conservation. As Carrier (2004) noted previously, the process of abstraction that seeps through discussions on the desirability of wolves in the Netherlands, in this case particularly and in environmental discourse in general, has become increasingly political. One of the underlying issues that triggers the discussion is the question which landscape should be taken as a benchmark for contemporary nature development; indeed, what is the threshold for "indigenousness" (cf. Martin and Trigger 2015) and who may decide this? Thirdly, the discussion about whether wolves should be regarded as native to the Netherlands also points to the increasing porosity of national borders, which relates paradoxically to the idea of nature that can be managed and created at will. Put differently, I suggest in this paper that the possibly trivial controversy of the *Wanderwolf* – indeed, for some it offered mostly hilarity – opens up a window on how global issues, such as climate change and the still growing group of refugees worldwide, enter local politics and concerns and are interpreted in multiple ways. Furthermore, following Hastrup (2014, 17), the *Wanderwolf* shows us, as anthropologists, "the challenge of integrating the wild and unexpected" in our analyses, "in the same manner as all people have to incorporate ferocious [...] facts into their daily lives". As Hastrup continues, "human history is always made within a complex framework of knowing, understanding, and acting upon the world – not only as it presents itself but as it is interpreted, tested, dealt with and imagined" (2014, 14). The influence of imagination in testing and contesting a particular understanding of the environment reveals precisely the increasing entanglement of the natural and human in the contemporary world, thus, disrupting accepted orders of control and agency.

A wolf in the Dutch Serengeti?

The media commotion surrounding the sightings of the *Wanderwolf* should be seen in the context of a much larger interest and effort to 'rewild' Europe (Rewilding Europe 2018). A particularly relevant starting point for the Dutch situation is the experimental design of a new ecosystem that has been taking place since the late 1980s in one specific area known as the Oostvaarders-plassen. The latter presents "an unlikely icon of European wildness" (Lorimer 2015, 97) and signifies a ground-breaking project in the Netherlands in an effort to recreate a particular version of primeval wilderness, sometimes rather courageously labelled "the Dutch Serengeti" (Tonnaer 2014). Under the gui-dance of Frans Vera, a rather controversial paleoecologist and conservationist, the area became part of an unconventional restoration strategy known, in rather oxymoronic terms, as "near-natural land management" or "nature development" (Neumann 1998, 28).

The national park, managed by *Staatsbosbeheer*, the Dutch Forestry Com-mission, is a nature reserve of 6 by 10 kilometres, northeast of Amsterdam. It is a vast marshland area – at least according to Dutch standards – populated by large grazers, such as red deer, Konik horses, and Heck cattle, and inhabited by tens of thousands of grey geese. The area was reclaimed from the sea in the 1960s and initially allocated for industrial purposes. When these developments did not transpire because of economic and hydraulic reasons (Lorimer 2015, 97), the Oostvaardersplassen is below sea level and requires continuous drai-nage, the abandoned area was increasingly colonized by wildlife. When Frans Vera took over the management of the area, he began to develop "an alter-native understanding of the paleoecology of Europe during the Holocene" (Lorimer 2015, 97), which informed his strategy for conservation. So-called *Pleistocene rewilding* takes as its benchmark a natural state prior to a sedentary human presence, suggesting that the prehistoric landscape was not a closed-canopy forest but rather a shifting mosaic of forest-pasture landscape kept open by large herbivores.

By reaching beyond a premodern agriculture to a deeper past, protagonists of the concept "have begun to erode the notion that historical dates linked to colonial contact are the obvious go-to baselines" (Marris 2011, 58). More-over, due to the many extinctions that have occurred during the past 13,000 years, the species used to recreate the landscapes can only be 'proxies', as a result of which it may be argued that, in fact, brand new ecosystems are being designed (Marris 2011). Through their grazing, the proxy animals, such as the Heck cattle and Konik horses, have indeed contributed to a specific natural dynamic, creating a mosaic of open water and marshland vegetation, inviting other spectacular flora and fauna species, such as the sea eagle and the raven, to settle permanently in the area.

In spite of its apparent ecological success, the idea of Pleistocene rewilding has been debated. Authors such as Lorimer (2015) and Marris (2011) focus particularly on the ecological science behind the experiment itself. I suggest here that the *Wanderwolf* introduced the next chapter in the rewilding design of the (Dutch) landscape, in so doing, inviting an enlarged socio-cultural view of the ambiguities of contemporary human–nature relationships.[4] More to the point, Pleistocene rewilding is not only divisive among ecologists and conservationists, but certainly also among the (lay) public. Rewilding excites but also divides equally as much. Opponents argue that the human presence and historically interwoven relationship between man and nature are intrinsic to Dutch natural and cultural heritage. In their view, new wilderness amounts to a largely virtual concept in a man-made country like the Netherlands and, thus, a manifest denial of a regionally developed attachment between man and nature.

The role the wolf plays in this disagreement is a complex one. As one of the potential top predators in Western Europe, the wolf has, according to rewilding theory, a key position to take up in co-creating a resilient and healthy ecosystem in areas such as the Oostvaardersplassen. Therefore, by developing the area further, 'the stage' is set for the return of the wolf. Vera, as quoted in Marris (2011, 66), is aware of the political minefield a reintroduction of the wolf will expose "in a country in which livestock owners are such a powerful political force". Already, death in the park – due to food shortages, for instance – has led to regular debates in the media about the artificiality of the ecosystem and animal cruelty as a consequence. Vera, however, accepts for now "the squeamishness" about death in wild ecosystems and instead, "he will just wait for wolves to come to him as they spread out from populations in Central and Eastern Europe. [...] 'They will come, whether we like it or not'" (Marris 2011, 66).

A public rehearsal for the return of the wolf?

Hitherto, no wolf has made it yet to the Oostvaardersplassen.[5] Even so, set against the background sketched above, when the *Wanderwolf* crossed the Dutch border in March 2015, it may have, unbeknownst, proclaimed the success of the rewilding movement. Predictably, the *Wanderwolf* was instantly incorporated as an iconic, incontestable furred PR body by Dutch organizations closely related to the rewilding movement. Nature conservationists in favour of the new wilderness were quick to initiate all kinds of educational and informational material, such as the platform *Wolven in Nederland* mentioned previously.

The information provided takes on a twofold approach, which Drenthen (2015) summarizes as "the wolf management" perspective. Firstly, in line with the aims of rewilding nature, the wolf's recovery is praised as a sign of ecological recuperation in the Netherlands. It is stated that the predatory behaviour of the wolf will add to the growth of even more biodiversity and help towards an ecological equilibrium. The return of the wolf confirms the development of "robust nature" and, simultaneously, creates it. Moreover, the information refers to models that present estimations that not only is there enough territory for the species to find food, the Netherlands may also, rather optimistically, be able to host several packs of wolves. Because the wolf is a 'synanthrope', it is likely that it will also ultimately settle permanently in the Netherlands. As the website of *Wolven in Nederland* explains: there are not many barriers, such as roads or rivers, that could thwart a wolf's path; wolves are experienced migrants and excellent swimmers. The wolf will ultimately decide for itself whether it wants to settle in our country (Wolven in Nederland 2018a).

The majority of the information is, however, as a second line, directed at deflating what ecologists term the Little Red Riding Hood syndrome, i.e., the variety of fables that circulate about the predatory and bloodthirsty nature of the animal. Instead, in their representation, the wolf is characterized as shy, intelligent and possibly somewhat lazy. In this kind of description, the wolf still takes on anthropomorphic faculties but rather than devouring young girls, old grandmothers or sheep, it is actually praised for its cleverness and agency to go about its way despite the overwhelming presence of humans. Drenthen rightly notes that "this strategy of normalization" entails a double standard of the role of fear: on the one hand, we are informed that humans have nothing to fear from the wolf, yet, at same time, in its position as the top predator in the ecosystem, fear of wolves is "an essential part of their functioning" (2015, 327).

Even so, if anything needs protection, it is the wolf itself, according to its proponents. The recently developed wolf protocol for the so-called first phase resettlement of the wolf, i.e. the incidental, wandering wolf, shows how even if the wolf wanted to live in anonymity, it will be identified, tagged and monitored intensely. The story about the *Wanderwolf* underlines this. It is known from which pack it originated, what sites it had visited, where it returned and how it died. Visitors to the website of *Wolven in Nederland* can, moreover, read in detail about "the return of the *Wanderwolf*" to the Netherlands (Wolven in Nederland 2018b). It is described how a team of the similarly named organization travelled to Hannover to pick up the frozen wolf, which was wrapped in foil and icepacks. The website states: "We were only allowed to leave after we had promised that the wolf would be mounted in a natural pose and not in an unnaturally aggressive posture. We solemnly promised this. After all, we received it for educational purposes" (author's translation). The *Wanderwolf* may be a lone wolf but as a true fairy tale celebrity it was immediately surrounded by a management team to be added to the spectacular wild.

Natuurmonumenten (literally: Nature Monuments), which is another foundation for the protection of Dutch nature and related to the platform *Wolven in Nederland*, regards the return of the wolf as "an enrichment for Dutch nature" (Natuurmonumenten 2018). It is assumed that it will fulfil the desire of many who long to see a wolf in the wild. However, enrichment may also be taken quite literally here. The wolf is not just yielding a profit for nature; there is money also to be gained for the catering industry and nature parks' administrators. Put differently, rewilding sells. A clear example of this has been the feature film documentary on the Oostvaardersplassen, entitled *De Nieuwe Wildernis* (2013), which was shown in the cinemas in late 2013.[6] The film quickly reached a platinum status thanks to more than 600,000 cinemagoers and was awarded several Dutch film awards. The success of the documentary demonstrated and contributed to a rekindling of public and mediatized attention regarding the Dutch landscape, even though it was also fiercely criticized for sanitizing the perspective from any human presence and the controversies surrounding the ecological management of the area.[7]

'The trouble with wilderness'

The visions expressed by the enthusiasts on the apparent return of the wolf reflect strongly those which Cronon (1995) has critically analysed in his influential essay on the notion of wilderness. He writes that the wilderness concept could become so pervasive and influential in the course of nineteenth and twentieth centuries because it got "loaded with some of the deepest core values of culture that created and idealized it" to such an extent that it became sacred (Cronon 1995, 73). Cronon points out, however, that there is "nothing natural about the concept of wilderness [...] a product of the very history it seeks to deny" (Cronon 1995, 79). This then, he writes, is the central paradox: "wilderness embodies a dualistic vision in which the human is entirely outside the natural". If true nature must also be wild, which the naturalized sacredness of wilderness suggests, then "our very presence in nature represents its fall" (Cronon 1995, 80–81).

There is no doubt that the great appeal of rewilded areas, such as the Oostvaardersplassen, is its seemingly 'unspoiled' character. Indeed, Frans Vera praises his creation as a "natural-processes-driven landscape"; natural is, in his view, better than cultivated (Marris 2011, 70). Marris marvels at the apparent paradox in his reasoning, for the Oostvaardersplassen is entirely cultivated and man-made. Yet, Vera is quoted saying that "the only thing man did was create the conditions and nature filled it in" (Marris 2011, 71).

As I have written elsewhere (Tonnaer 2014), the human-made history of the park and the continuous interventions needed to keep it 'wild' both ecologically and socio-culturally, represents a continuation of the colonial genesis of many national parks across the globe.[8] The historical development of national parks, such as the famous Serengeti National Park, hinged largely on imaginaries in which a romanticized wilderness was opposed to the decadent metropole (Neumann 1998, 18). As noted by Cronon (1995), and Salazar (2006) for East Africa, national parks and wildlife conservation became closely tied to "a peculiarly bourgeois form of antimodernism" (Cronon 1995, 78). Put differently, for the (colonial) elite, wilderness became something to observe and enjoy as visitors rather than as inhabitants. The emergence of safari tourism in Africa needs to be understood in this historical light (Neumann 1998; Salazar 2006) but may also be extended to contemporary linkages between nature conservation and tourism (see Salazar and Graburn 2014).

Furthermore, according to Neumann (1998, 9), the "national park ideal" – a particular conceptualization of nature that, since the foundation of the renowned Yellowstone National Park in the United States, has served as a model for national parks worldwide – reflects the "landscape way of seeing: the removal of all evidence of human labor, the separation of the observer from the land, and the spatial division of production and consumption. *A national park is the quintessential landscape of consumption for modern society*" (Neumann 1998, 24; emphasis in original). In a similar vein, Carrier (2004, 6) points out that nature and the environment are construed in the environmentalist movement as a realm that is separate from social practices and human experience. Accordingly, the value of an experience of the wilderness lies in a kind of 'placelessness', of being the Other in the landscape.

There is no univocal agreement about the place of visitors in the rewilded landscapes in the Netherlands. Some of the stricter ecologists prefer to minimize human presence, urging to close rather than open nature reserves for recreation, turning the experience of wilderness even more into a privileged opportunity for those who have the authorization to enter an area and are capable of recognizing its natural values. Vera's choice of Heck cattle and Konik horses in the Oost-vaardersplassen, for instance, was deliberate; he wanted the public to see them as wild animals and to prevent a line-up of girls along the fence in the case of more cuddly looking horses (Marris 2011, 59), implicitly suggesting that there are 'right' and 'wrong' ways of relating to and being in nature.

The question of who or what the visitor is and who or what the dweller in the natural landscape became exacerbated when the *Wanderwolf* entered the country. The promise of spotting wild animals as an exciting addition to one's leisurely stroll through nature was certainly not adopted by all Dutch. The debate on the desirability of the return of the wolf reflected in essence that which Cronon notes as the actual trouble with wilderness, i.e. the inherent dualism in the concept of wilderness encourages "its advocates to conceive of its protection as a crude conflict between the 'human' and the 'nonhuman' – or, more often, between those who value the nonhuman and those who do not" (Cronon 1995, 85). Indeed, opposition against the enthusiasm of the wolf's welcomers turned on the issue of 'sparing' or 'sharing' nature (Lorimer 2015).

'Nature stupidity'

The wolf's adversaries spoke out equally during the *Wanderwolf*'s appearance. After the wolf was first sighted, the obvious organizations were consulted, such as the National Association of Professional Sheep Farmers and the National Hunting Organisation, to give their comments. They voiced their concern for their stock and possible financial costs they would incur in the future from the loss of sheep. More interestingly though were the responses of the lay public to the news of the *Wanderwolf*. Some of the views resembled those described elsewhere (Skogen, Mauz, and Krange 2008), for instance, suggesting that the *Wanderwolf* was deliberately released in the area by ecological fanatics or that the wolf was, in fact, not a real wolf but a particular species of dog.

The majority of the responses, however, accepted the validity of the wolf claim, i.e., the wolf was a real wolf, but questioned the desirability and wisdom of welcoming the wolf onto Dutch territory. Notwithstanding the reassurance of ecologists about the wolf's shyness of humans, there was heightened concern for one's safety. In response to an article that appeared in one of the weekly newsmagazines in the same week that the *Wanderwolf* entered the country, one of the comments stated: "imagine walking with your kids in the forest, suddenly such a beast is standing in front of you. You will not be happy, take my word for it". And another wrote: "I find it completely over the top to be happy with the wolf. Those people also want to see bears here? Imagine walking with your dog and a wolf crosses your path" (Comments Section 2015; author's translation).

However, the majority of the negative responses indicated a much more fundamental doubt about the existence of Dutch wilderness. The wolf was accordingly described as a tourist who had not read the travel brochure well and as an 'ecological refugee', for whom the asylum procedure had already begun. In short, apart from fear, there was widespread scepticism about whether there is enough space for the wolf to actually prosper. In the wording of one on the website mentioned above:

> We live with 17 million people in a very small country, belonging to one of the most densely populated countries in the world! The Netherlands no longer has any nature; merely a few forest parks. No packs of wolves can live there without big problems. Look at that little overpopulated park in Flevopolder [the Oostvaardersplassen] where horses and deer perish [due to a lack of food]. We do not have any nature left in the Netherlands for straying packs of wolves. This is nature stupidity.

Criticism on the dominance of the rewilding ecologists, "little green eco-tyrants" as one commentator termed them, and their enthusiasm for the wolf reveals the disparagement of the concept of new wilderness in the Netherlands. The promised enjoyment of the spectacular scenery of wilderness and, hence, possibly the wolf almost inevitably entails a disregard of regionally developed attachments between man and nature. Returning the land to nature – 'sparing' rather than 'sharing' if only as visitors – deprives, in the view of its antagonists, the country of regional diversity, the cultural-historical legacy and 'legibility' of the land, and, accordingly, a sense of place (Drenthen 2009, 212). The *Wanderwolf*, itself unaware of its impact, thus signalled the earnestness of contemporary complexities in human–nature entanglements.

Bewildering entanglements

Wolves are powerful symbols of wilderness, because, as Skogen, Mauz, and Krange (2008, 125) write: "if they [wolves] can recover, wild nature may not be doomed after all" (see also Van der Meulen 2016). Even though the

Wanderwolf was in actual hairy fact present, the landscape into which it wandered is highly contested in terms of its management. Following Tsing's (2014, 36) emphasis on recognizing "the entangling of multiple scales and trajectories in the making of social landscapes", the story of the *Wanderwolf* invites us to consider the multilayered and often ambiguous human–nature sociality in contemporary society. The *Wanderwolf*, I suggest, points not only to unstable categories of 'hosts' and 'guests' but also shows that nature is always implicit in the making of social life (Hastrup 2014). Therefore, the return of the wild inevitably opens up a plethora of questions that implicates much more than the question of the 'right' ecological management and conservation.

To begin with, a neoliberal logic can be seen to emerge in the keen embracing of the *Wanderwolf*. The Dutch environment seems to be increasingly shaped and valued as a malleable, spectacular landscape. In the discourse on the wolf's return and, hence, its apparent corroboration of successful landscape management, nature has turned into a commodity that is accordingly submitted to the principles of the market, as can be seen in references to notions of gains, losses, consumption and production that arose when the *Wanderwolf* entered the country. In that regard, the rewilding movement skilfully played on the possibilities the market mechanism seems to invite. Whereas the financial budget and influence of the rewilders is still growing, smaller nature organizations advocating for recognizing the cultural history embedded in nature, including the nature knowledge of farmers (De Stoop 2015), are increasingly pushed to the margins and stigmatized as practitioners of destructive methods of landscape management. The economic and political ability of the wolf's protagonists to determine the dominant view of the Dutch landscape is, thus, also growing.

Secondly and related to this, the discrepancy between experts and the public in terms of authority is widening. As Carrier notes,

> the environmentalist movement has generated a mass of arguments to make conservation appear important. Perhaps reflecting the fact that 'the natural environment' and 'nature' are abstract concepts, overwhelmingly these arguments are abstract and general. That is, they reflect and embody a logic and set of assumptions that tend to transcend particular places and times.
>
> (Carrier 2004, 5)

Debates about the desirability of wolves in the Netherlands are dominated by ambiguities surrounding notions such as sustainability and biodiversity, compound terms that have little imaginative weight for the lay public and that, furthermore, deny "significance to (...) individual and local experiences" (Carrier 2004, 8). This process of abstraction is, therefore, according to Carrier, also political.

A particularly knotty issue arising from this is: which landscape should be taken as the benchmark for contemporary nature development? In the view of critiques of the new wilderness, current processes amount to nothing more than "nature forgery" (De Stoop 2015, 227). Indeed, protagonists of rewilding do not seem to deny this; the apparent leap that they have made is admitting that we already live in an intensely managed world (Marris 2011, 64). Even so, underlying the new wilderness ideology is the culturally pervasive, yet very modern model that is governed by the principle that the greater the temporal distance between one's historical benchmark and the here and now, the more authentic one's acclaimed 'natural landscape' is. Indeed, the authenticity claim and, in its wake, the decisive authority to intervene in the landscape made by the rewilding protagonists rests on the idea that the time travel involved in seeing wilderness is larger and, therefore, more primitive and purer than the cultural historical polders and farm-lands. But what is the threshold for indigenousness (cf. Martin and Trigger 2015)?

Moreover, as Cronon has already noted, the crude conflict between 'human' and 'nonhuman', which is the dualism that the notion of wilderness intrinsically holds, also tends to ignore crucial differences between humans and "the complex cultural and historical reasons why different peoples may feel very differently about the meaning of wilderness" (Cronon 1995, 85). Australian Aborigines, for example, perceive 'wild country', empty land that from a European perspective appears as wilderness and, therefore, as 'natural', as land that is marked by loss, a loss of life-support systems and of relationships among living beings and their country (Rose 1996, 20–21; Tonnaer 2016, 2019). Closer to home, the increasing devaluation of productive labour and concrete knowledge that grows from working the land with one's own hands, a knowledge often deeply etched into the bodies across generations, is having upsetting effects on the lives of many rural people. Indeed, De Stoop writes in his non-fictional book on another, controversial nature area near the Dutch–Belgium border, that farmers see themselves turning into the Aborigines of the West: dispossessed and driven away from their land, having their values and connections to land not recognized (De Stoop 2015, 276).

Finally, the discussion regarding whether the wolf may or may not be regarded as native to the Netherlands also connects to international debates both on the current large-scale migration and uprootedness of many people and the profound concerns regarding climate change. For the first, it may be noted that the way in which the wolf was being discussed, both by its supporters and its opponents, reveals uncanny discursive similarities with the equally heated debates on how to deal with the still-growing group of refugees that seek asylum in Europe, and who have become the plaything in EU political trade-offs. The mobility of wolves throughout Europe and the EU conventions that protect the wolf and regulate its behaviour at the same time show the porosity of national borders in multiple ways. Its agency as a very mobile animal, known for traversing large distances, is, in that regard, delicately ambiguous. The news that the wolf will decide for itself whether it wants to come to the Netherlands is, thus, perceived both positively and negatively. Indeed, both the wolf and migrants seem, obviously for very different reasons, not to adhere to the "national order of things" (Malkki 1995).

The adversaries of the wolf's return make visible the uncertainties that many people experience in these mobile times, living also under the cloud of the Anthropocene, under which many perceive the worrying news on and experience the calamitous consequences of climate change. Such uncertainty sparks an interest in a cultivation and revitalization of the regional and the local, and the cultural and historical particularities of place and identity (Olwig and Jones 2008; Simon, Huigen, and Groote 2010). I second Carrier's point when he writes that "our attention to local contexts, then, needs to be complemented by an awareness of their relationship with larger contexts, and hence indirectly with other local contexts" (2004, 14). The right to belong, therefore, does not merely affect the wolf but transcends the animal domain to include human social and cultural processes as part of the all-encompassing environment.

Indeed, signalizing the difference between controlled rewilding and spontaneous rewilding (Drenthen 2015, 321), the wolf shows how nonhumans have their own agency and autonomously yet effectively may intervene in the (history of) nature development that always implicates the human world. Following Tsing (2014, 33) we, as humans, may not always be in charge and we might, therefore, want to learn about the social worlds that other species help to build.

As media announced during the *Wanderwolf* incident, the comeback of the wolf is no longer a fairy tale. On the contrary, I would argue that this is a new story in the making about which we cannot easily tell, like in so many other proper tales, whether the moral is right or wrong.

Epilogue

Since the *Wanderwolf*'s visit in 2015, there have been many more sightings of wolves. The number of wolves in the border region has also clearly been increasing. In September 2016, a second wolf, post *Wanderwolf*, was spotted in the same region. The media outburst that I describe in this paper for the *Wanderwolf* did not occur to the same degree for the later arrival. Nonetheless, each new sighting of a wolf in the country is still considered newsworthy. At the end of 2018, *Wolven in Nederland* stated that wolves have, in their definition, unquestionably returned to the Netherlands, as, at that point, there had been a continuous presence of one or more wolves on Dutch ground for more than a year and a half. The news of 2019 was that wolf cubs had been born in the Netherlands; however, this news has not received any positive follow-up since.

In 2018, the *Wanderwolf* itself was on tour again, as part of an educational exhibition and displayed in a centre for sustainability in Assen in the province of Drenthe. The exposition was both on the history of the wolf in the country and its biological features and ecological functions. Visitors could see the skeleton and skin and learn, moreover, about the process of stuffing the dead animal. What the exhibition showed above all is how, for better or worse, the *Wanderwolf* has been adopted as an icon of the changing Dutch landscape. The *Wanderwolf* may no longer actively roam the land, however, in its afterlife it may still have a prominent part in the shaping of it.

Notes

1 A note on spelling is needed here. *"Wanderwolf"* is the German name for the wolf, but it is also the name under which the animal has become known predominantly in the Netherlands. *Wandelwolf* is the Dutch name. *Wandelen* in Dutch means walking but also carries the connotation of wandering. *Wandern* in German has a similar meaning. Both names refer to the long distance this particular wolf traversed through Germany and the Netherlands seemingly without a particular direction in mind. Ecologists, however, presume that the young animal was in fact looking for a territory to make his own. I follow the spelling that is now consistently used to refer to the wolf and, thus, write its name with a capital "W".
2 The organization *Wolven in Nederland* is a collaborative initiative of several nature conservation organizations that strive for a conflict-avoiding coexistence between wolves and people.
3 This wolf has become known as the 'wolf of Luttelgeest'. It has been stuffed and is kept at the Museum Naturalis in Leiden.
4 Although in stating this I am not doing proper justice to both studies that are thorough, fascinating and thought-provoking in more domains than merely the analysis of the ecological system.
5 When almost half of the park's animals died because of starvation in the winter of 2018, protests from the public became so serious and enraged that since then, management of the park has taken a different course. This has led, among other things, to a move of a significant number of animals to nature areas elsewhere in Europe. Some ecological experts have denounced the park as a 'failed experiment'.
6 http://emsfilms.com/de-nieuwe-wildernis/ [accessed 17 December 2019].
7 Indeed, parodies as well as a short contra-documentary have been made in response to what is considered an invented spectacle of wild nature with disastrous consequences for the nature that existed there previously (see e.g. Schaper 2018).
8 The allusions to the African Serengeti National Park were prominent in the motion picture mentioned above.

Bibliography

Carrier, J. G. 2004. "Introduction." In *Confronting Environments: Local Understandings in a Globalizing World*, edited by J. G. Carrier, 1–29. Walnut Creek, CA: Altamira Press.
Cronon, W. 1995. "The Trouble with Wilderness; or, Getting Back to the Wrong Nature." In *Uncommon Ground: Rethinking the Human Place in Nature*, edited by W. Cronon, 69–90. New York, NY: W. W. Norton & Company.
Comments Section, 2015. "De Wolf is Terug in Nederland en deze Foto bewijst het." *Elsevier Weekblad*. 2016. http://www.elsevier.nl/Nederland/achtergrond/2015/3/De-wolf-is-terug-in-Nederland-En-deze-foto-bewijst-het-172129W [accessed 7 July 2016].
De Nieuwe Wildernis, 2013. "De Nieuwe Wildernis: Grote Natuur in een Klein Land." http://emsfilms.com/de-nieuwe-wildernis/ [accessed 31 January 2018].
De Stoop, C. 2015. *Dit is Mijn Hof*. Amsterdam: De Bezige Bij.
Drenthen, M. 2009. "Developing Nature along Dutch Rivers: Place or Non-Place." In *New Visions of Nature: Complexity and Authenticity*, edited by M. Drenthen, J. Keulartz, and J. Proctor, 205–228. Dordrecht: Springer.
Drenthen, M. 2015. "The Return of the Wild in the Anthropocene. Wolf Resurgence in the Netherlands." *Ethics, Policy & Environment* 18, 318–337.

Hastrup, K. 2014. "Nature: Anthropology on the Edge." In *Anthropology and Nature*, edited by K. Hastrup, 1–26. New York, NY: Routledge.

Lorimer, J. 2015. *Wildlife in the Anthropocene: Conservation after Nature*. Minneapolis, MN: University of Minneapolis Press.

Malkki, L. H. 1995. "Refugees and Exile: From 'Refugee Studies' to the National Order of Things." *Annual Review of Anthropology* 24, 495–523.

Marris, E. 2011. *Rambunctious Garden: Saving Nature in a Post-Wild World*. New York, NY: Bloomsbury.

Martin, R. J. and D. S. Trigger. 2015. "Negotiating Belonging: Plants, People, and Indigeneity in Northern Australia." *Journal of the Royal Anthropological Institute* 21, 276–295.

Natuurmonumenten, 2018. "De Wolf." https://www.natuurmonumenten.nl/standp unten/wolf [accessed 31 January 2018].

Neumann, R. P. 1998. *Imposing Wilderness: Struggles over Livelihood and Nature*. Berkeley,CA: University of California Press.

Olwig, K. R. and M. Jones. 2008. "Introduction: Thinking Landscape and Regional Belonging on the Northern Edge of Europe." In *Nordic Landscapes: Region and Belonging on the Northern Edge of Europe*, edited by M. Jones and K. R. Olwig, ix–xxix. Minneapolis, MN: University of Minnesota Press.

Rewilding Europe, 2018. "Rewilding Europe: Making Europe a Wilder Place." https:// www.rewildingeurope.com/ [accessed 31 January 2018].

Rose, D. B. 1996. *Nourishing Terrains: Australian Views of Landscape and Wilderness*. Canberra: Australian Heritage Commission.

Salazar, N. B. 2006. "Touristifying Tanzania: Local Guides, Global Discourse." *Annals of Tourism Research* 33, 833–852.

Salazar, N. B. and N. H. H. Graburn. 2014. "Introduction: Toward an Anthropology of Tourism Imaginaries." In *Tourism Imaginaries: Anthropological Approaches*, edited by N. B. Salazar and N. H. H. Graburn, 1–25. New York, NY: Berghahn.

Schaper, S. 2018. De Nieuwe Wildernix. https://www.youtube.com/watch?v= FWvFAGWwmeg [accessed 31 January 2018].

Simon, C., P. Huigen, and P. Groote. 2010. "Analysing Regional Identities in the Netherlands." *Tijdschrift voor Economische en Sociale Geografie* 101, 409–421.

Skogen, K., I. Mauz, and O. Krange. 2008. "Cry Wolf!: Narratives of Wolf Recovery in France and Norway." *Rural Sociology* 73, 105–133.

Tonnaer, A. 2014. "Envisioning the Dutch Serengeti: An Exploration of Touristic Imaginings of the Wild in the Netherlands." In *Tourism Imaginaries: Anthropological Approaches*, edited by N. B. Salazar and N. H. H. Graburn, 242–259. New York, NY: Berghahn.

Tonnaer, A. 2016. "Intersecting Journeys of Past and Present in the 'Bush': Unsettling Coevalness in the Tourist Space of Indigenous Australia." *International Journal of Tourism Anthropology* 5, 172–186.

Tonnaer, A. 2019. "On the Beaten Track: Ambiguous Wilderness in the Tourist Space of Indigenous Australia.", Chapter 8 in *Ecocritical Concerns and the Australian Continent*, edited by B. Neumeier and H. Tiffin. Lanham, MD: Lexington Books.

Tsing, A. 2014. "More-than-Human Sociality: A Call for Critical Description." In *Anthropology and Nature*, edited by K. Hastrup, 27–42. New York, NY: Routledge.

Van der Meulen, D. 2016. *De Kinderen van de Nacht: Over Wolven en Mensen.* Amsterdam: Em. Querido's Uitgeverij BV.

Wolven in Nederland2018a. "Kunnen wolven in Nederland leven?" https://www.wol veninnederland.nl/#main [accessed 31 January 2018].

Wolven in Nederland2018b. "Wandelwolf terug in Nederland." https://www.wolve ninnederland.nl/nieuws/wandelwolf-terug-nederland [accessed 31 January 2018].

Zwart, H. 2003. "Aquaphobia, Tulipmania, Biophilia: A Moral Geography of the Dutch Landscape." *Environmental Values* 12, 107–128.

5   "One feels a shiver" – wolf perceptions and representations in Portugal

Filipa Costa

For two years I have lived and worked in a camping site in the extreme north of Peneda Geres National Park (PGNP), where livestock graze freely in the mountains and wolves have always existed. During this period I hosted park visitors and learned about their expectations. I have also immersed myself in local social life, participating in social events and daily routines. I came to realize that the wolf is at the centre of an intricate social web where several agents weave different discourses and hold equally strong emotions about this elusive creature. I was once helping a family from Lisbon planning some activities when the mother surprised me: "Well, the kids want to see wolves". She said it so casually it made me wonder if the wolves pictured in conservation and touristic campaigns looked friendly enough to make people think they could be found easily. Certainly not everyone arrived with that kind of unrealistic expectation but the interest in the wolf was quite common among visitors, sometimes close to fascination. There were also those who came regularly with a mission: to find a wolf and to photograph it. These wolf enthusiasts were willing to spend some nights up in the mountains to catch that rare moment of encounter on camera. I found that to spot a wolf or to hear them howling is extremely exciting to tourists and locals alike. Locals and outsiders share their wolf stories, their own personal encounters, in social events, mentioning sometimes how afraid they felt, how they were in a hurry to leave. A respectful reverence and a sense of pride are implicit in their narrative. Others, particularly livestock owners, express different feelings often displaying despite. Depreciative comments, however, are not always expressed freely in front of outsiders and awkward silences sometimes fill the room when the wolf is the subject. While I lived there, a wolf was actually killed during a legal wild boar hunt, and although many of the locals knew about the circumstances and who did it, the information was preciously kept and never discussed with outsiders. There was this tense atmosphere for quite some time; a big, white and spacious elephant-wolf in the room. The different discourses about the wolf that coexist and sometimes collide in the local context I have briefly described have, nevertheless, one thing in common: the tendency to avoid locating where a wolf was seen, to hide the location like a trophy as if the wolves should remain a secret to which access is a privilege, whether to kill, photograph or protect them.

This life experience in PGNP motivated me to research people's relationship with nature and the processes through which particular natural elements acquire a special status, namely, a heritage status. I have lived a total of five years in the region, two of them in the context described above – managing a camping site – which placed me in a privileged position interacting with visitors and locals alike. This chapter is inspired by that personal experience. The data about local wolf perception mentioned here were collected during that period. I have also, at the same time, gathered information about visitors' perceptions of nature and the wolf. Here, I am confronting those data with relevant literature and complementing them with data from my ongoing PhD research about natural heritage and tourism in PGNP, where I use the wolf as a case study to analyse how certain natural elements acquire a heritage status. I am currently conducting fieldwork with park visitors and accompanying different groups in guided pedestrian walks to explore different perceptions of nature and natural elements. At the same time, I am analysing how nature and selected natural elements are represented in the park discourses through an analysis of institutional maps, exhibitions, texts and images. Concerning the wolf, I am also devoting attention to touristic and environmental discourses in a less localized fashion by following media and social media and attending related events, such as conferences and exhibitions. Thus, referring to this particular chapter, the data being used here come from three sources: 1) a localized personal experience in a specific PNPG area that allowed me insight into locals and visitors' perceptions of wolves; 2) my ongoing fieldwork with park visitors where I am using participant observation as a method to gather more information about people's perceptions of the wolf; and 3) my ongoing research about wolf-related touristic and environmental discourses on a delocalized level not limited by the PGNP boundaries.

The wolf is precisely one of those elements I am interested in that have become more than a creature: it is an icon for a specific kind of nature. In the present chapter, I aim to reflect on different perceptions of the wolf and, for analytical purposes, I simplistically separate them into the traditional and ecological, while exploring concepts closely linked to the status the wolf has achieved. Generally speaking, the traditional wolf is the one described as a shepherd's enemy, illegally killed in secret, and the ecological one is supposed to replace it as an essential part of a balanced ecosystem. I argue that these perceptions of the wolf are not opposites, because they feed on each other. The local traditional wolf is as much revered as the delocalized ecological one is constructed and anthropomorphized. I suggest the wolf's iconic status can be translated as heritage, in the Portuguese context, and that the wolf that achieved such a status represents, in fact, a combination of these two perceptions into a creature I call the noble savage. I use this concept quite colloquially without going further into a historical and philosophical dissection. I mean something that is wild in its nature – free of human touch – but imminently good in its essence.

In the first part, I explain why the wolf can be understood as a cultural keystone species in Portugal, considering the concepts of heritage and endangerment. In the second part, I present the Iberian wolf and its context and then discuss the ambiguous traditional perception of this species, illustrating the existing research with my own data. The ecological wolf is the subject of the third part, where I inevitably explore the wilderness concept and the current paradigm of nature conservation, questioning then the apparent neutrality of the way the wolf is presented as an ecological entity. In the conclusion, I introduce the noble savage or the wolf heritage and argue that this is the one that actually dominates the perception and representation of the species in the Portuguese context.

The wolf: a cultural keystone species

Human–wolf coexistence throughout Europe has always been hard but today the wolf's iconic status as a symbol of wilderness is aiding conservation efforts. For Garry Marvin, the wolf "carries the cultural weight of being emblematic of the wilderness a necessary physical and spiritual antidote to modern material civilization and a symbol of the well-being of that wilderness" (Marvin 2012, 159). Although wolves were eradicated from a great part of Europe by the nineteenth century (Marvin 2012, 83), an autochthonous subspecies survived in the Iberian Peninsula. The Iberian wolf (*Canis lupus signatus*), a subspecies of the European grey wolf (*Canis Lupus*), is categorized as endangered in Portugal. It perseveres despite residual persecution, occupying a privileged place in the emblematic species pantheon. The ancestral wolf–shepherd conflict perseveres as well but is being confronted with the species' growing popularity. Perceived as a charismatic survivor, the wolf is today a valuable element of the national natural heritage.

In November 2016, in a seminar entitled Conserving Large Carnivores in a Crowded Europe, organized by CIBIO – Research Centre in Biodiversity and Genetic Resources, the zoologist Jonh Linnell started his presentation by asking what the real motivation to study wolves is. He answered himself: "These are spectacular animals. They are cool. They do something to us." The question that remains unasked, however, is: why? In fact the wolf has not always been such a charismatic figure. Marvin analyses how its cultural image has been continuously altered, demonstrating that the way we conceptualize species and nature itself keeps changing. In 1926, the last Yellowstone wolves were killed by the park guards as they were then considered creatures to exterminate (Marvin 2012, 75). William Hornaday, a conservation pioneer, described them as "degenerate" and "unmoral" (Marvin 2012, 115), and it was not until the 1930s that wolves became part of the natural balance, with intrinsic value, very much due to the cultural re-evaluation introduced by the Endangered Species Act (Marvin 2012, 118). Nevertheless, to be understood as part of an ecosystem by scientists does not equal entering the realm of admired animals. The concept of heritage is useful to understand at what point something gains increased value in the public eyes. Agreeing with Rodney Harrison, I will understand heritage not as a passive way of preserving elements from the past, but as "an active process of assembling a series of objects, places and practices that we choose to hold up as a mirror to the present, associated with a particular set of values that we wish to take with us into the future" (Harrison 2013, 4). Although there is not much research about animals as heritage, Peter Coates (2015) has written a unique article on the subject. According to him, animals have entered heritage discourses since the middle of the nineteenth century as "objects of heritage that are cultural as well as environmental" (Coates 2015, 289). Nowadays, the perception of animals as heritage implies that "cultural and national particularities are displayed as well a commitment to biodiversity" (Coates 2015, 276). Coates defends the concept of cultural keystone species, originally used for species significant to indigenous populations, and states that they also make sense in the Western world (Coates 2015, 296).

Species that belong to the "endangerment landscape" (Vidal and Dias 2016, 23) are particularly prone to seeing their value increase and their status rise in the heritage hierarchy. Thus, the "scarcity theory" (Coates 2015, 297) is fundamental to understanding why some species get more attention from the public. Being endangered is an essential aspect, because extinction and loss of diversity are "anti values par excellence" in the modern world (Vidal and Dias 2016, 1). Although some effort was put into preventing certain creatures' disappearance, during the nineteenth century, extinction was not considered as affecting the natural balance (Sepkoski 2016, 71). According to David Sepkoski, there was a paradigmatic change around the 1970s and 1980s when loss of biodiversity and species extinction started to be conceptualized in terms of a global crisis. Influenced by palaeontology, biology then considered extinction no longer as an inevitable natural selection process but as a phenomenon that can happen suddenly due to geological catastrophe. This paradigmatic change contributed to "a greatly enhanced sense of impending danger that was absent in earlier conservation rhetoric" (Sepkoski 2016, 78). Endangered entities – species, ecosystems, languages, artefacts, etc. – emerged in political and public discourses (Sepkoski 2016, 62). Since then, diversity has been seen as a precious resource with normative and absolute value (Sepkowki 2016; Vidal and Dias 2016). Fernando Vidal and Nélia Dias argue that precisely from the 1980s onwards, a feeling of constant threat has dangled over nature and culture. They have coined this feeling as endangerment sensibility and define it as part of a "complex of knowledge, values, affects and interests characterized by a particularly acute perception that some organisms and things are «under threat», and by a purposeful responsiveness to such a predicament" (Vidal and Dias 2016, 2).

However, not every endangered animal attracts attention and affection. How many of us could identify an endangered insect? Some emblematic species seem to effectively dominate the "endangerment landscape" (Vidal and Dias 2016, 23). In the 1960s, for example, the focus was on big mammals, such as the giant panda that has today been replaced by the polar bear, an Anthropocene symbol (Coates 2015, 275). According to William Cronon, endangered species "serve as vulnerable symbols of biological diversity while at the same time standing as surrogates for wilderness itself" (Cronon 1996, 82). That is indeed the case of the Iberian wolf, whose endangered condition places it in a fragile situation in need of human intervention: it stands as an icon for the auto-chthonous biodiversity that is threatened by invasive species and poor man-agement. Its role in the ecosystem is frequently reinforced as an irreplaceable piece of the natural diversity necessary to maintain a healthy balance between species. The wolf represents that very same nature that needs protection and safeguarding. In addition to this "commitment to biodiversity" implied in the devotion to the wolf, its heritage status and value also depend on its native character because "things indigenous are also more authentic" (Coates 2015, 291). The species autochthonism grants it legitimate nationalist value as heri-tage, as it guarantees its natural belonging. For those reasons, the wolf is defi-nitely a cultural keystone species. It is not only a creature but an element of heritage that people particularly cherish. In social media, it is described as "noble", as a "survival", as "authentic", "beautiful" and "magnificent". Visi-tors to the national park ask about it frequently and express how they would love to see one. Joan Frigolé states that the process through which the wolf is turned into heritage "is projecting onto this species the meanings and values of ecology and diversity" converting it into an "eco-symbol" (Frigolé 2010, 23). In the Portuguese context, I defend that the heritage status depends on more than this, because besides this ecological wolf, there is also the one that makes us shiver. So, which one of them does something to us and is so highly placed in the heritage ranks? For now, I will consider them separately.

The Iberian wolf or "the vermin and the noble beast"

At the beginning of the twentieth century, wolves were spread all over the country, but with more roads and fewer forests, prey diminished and habitat shrank. Together with continuous persecution, this resulted in the species approaching extinction. The Iberian wolf has been legally protected since 1988. The surviving populations are concentrated in the mountain territories in the north and centre in depopulated rural areas where traditional pastoral practices provide some prey for the only predator remaining. Nowadays, the wolf population is considered stable. The last census was carried out in 2002 and 2003 and estimated a population of 300 individuals divided into 65 packs (Álvares 2006). A conservation plan that includes producing a new census is currently awaiting governmental approval. This plan has the objective of improving human–wolf coexistence and promoting wolf conservation through the inclusion of local knowledge into conservation initiatives. It follows recent tendencies that defend a multidisciplinary approach to conflict, valuing social and cultural dimensions as a way to effectively change negative attitudes towards the species (Álvares et al. 2011; Boitani 2000; Dressel, Sandstrom, and Ericsson 2014; Lopes-Fernandes et al. 2016). It also stresses the advantages of wolf-related tourism to reduce conflict. Some conservationists defend that accepting the wolf as a touristic resource may improve local perception (Álvares 2006; Álvares et al. 2011; Boitani 2000). Throughout north Iberia, antique stone wolf pit structures – the *fojos* – can be found, some with large dimensions and strong impact in the landscape. The knowledge and practices associated with them are unique. This legacy "constitutes a unique opportunity to promote the recreational and utilitarian value of the wolf, not only as a key species in biodiversity but also as an important element in rural culture and tradition" (Álvares et al. 2011, 327). In PGNP – a territory of 700 km^2 in northwest Portugal where a stable wolf population contributes to individual dispersal (Álvares 2006) – there has been a gradual increase of wolf-related touristic offers in the last few years, such as guided visits to the *fojos* or interpretative tours to find wolf faeces or footprints, learning about their habits. The "wolf territory" is becoming a brand. As Marvin puts it, "The new devotees of lupophilia are not simply interested in reading about wolves but also want to experience them in the wild" (Marvin 2012, 164).

If tourism will improve the traditional human–wolf conflict remains to be seen, but this phenomenon demonstrates how the wolf is becoming increasingly popular. Every once in a while, a wolf killed in illegal traps or shot gets media attention. News about wolves allegedly attacking herds frequently gets to the public as well. Public reactions are very emotive in any of the cases. The wolf is traditionally seen as an enemy of the shepherds, but the feelings towards them vary between admiration and hatred, as ethnography and literature confirm (Álvares et al. 2011; Dressel, Sandstrom, and Ericsson 2014; Lopes-Fernandes et al. 2016; Marvin 2010). A very interesting article about the wolf in Portuguese literature analyses several occurrences and concludes that there is "an ancient rivalry with the super-predator, deeply rooted in symbolic, cultural, and psychological features" but also that different images coexist, as it is "simultaneously a voracious predator and a well-known creature admired for its strength and intelligence" (Lopes-Fernandes et al. 2016, 15). Thus, literary analysis demonstrates how the wolf is simultaneously seen as "the vermin and the noble beast" (Lopes-Fernandes et al. 2016, 5). Another research paper about attitudes towards the wolf in the Iberian Peninsula also mentions the coexistence of two perspectives closely connected: "one seeing the wolf as a real threat to livestock and the other considering this carnivore as a symbolic identity (…) implying that the wolf–human conflict is rooted not only on this carnivore's role as an enemy to livestock husbandry, but also in its mythical and supernatural essence" (Álvares et al. 2011, 315). Álvares et al. also suggest that "this ambiguous perception of Iberian communities coexisting with this carnivore may have allowed the continued survival of a wolf population in the Iberian Peninsula" (Álvares et al. 2011, 326).

My experience confirms this ambiguity. It is very common in the PGNP territories to hear that if one is alone in the mountains and there is a wolf around, even if it cannot be seen, one will "feel a shiver". I have heard this many times from older and younger people, some describing a personal experience, while others mentioning a common belief. Álvares et al. mention the same effect in a wider area and over time: "the proximity of a wolf causes one's hair to stand up or renders a person speechless for several hours or days", a belief "reported by Roman authors as early as the first century AD" (Álvares et al. 2011, 321). The presence of the creature is considered powerful enough to be felt even when it is out of sight. This suggests its supernatural character and the profound effect the wolf has on people. It suggests that the animal is indeed more than mere vermin or an evil opportunist. It confers on the wolf a mystic aura. There is another phenomenon worthy of mention: the "half wolf" dogs. It is very common in the PGNP to hear someone saying that he/she has or had a friend that owns a "half wolf" dog. Being a "half wolf" dog seems to be better than being just a dog. People mentioning it considered them bigger and better than normal dogs, owing their charisma to the wolf side. It does not really matter whether the animals mentioned were in facts hybrids (which I suspect they were not); what does matter is the value attributed to those animals. Even in contexts where the wolf is not appreciated, some of its characteristics are praised, at least in dogs. These data suggest that, in fact, hostility can coexist with a positive image of the wolf as a strong and fearless being that can contribute to breeding an admired kind of dog.

There is an extremely rich reservoir of beliefs about the wolf in Portugal that needs more research. It would be equally important to explore how perceptions of the wolf in rural contexts may be influenced by nature conservation programmes. There is a belief in the PGNP that can also be found in other European countries (Álvares et al. 2011) and that can actually be considered a modern myth: wolf release. I have frequently heard stories of wolf encounters in the past featuring the wolves "from before". These were described as being very different from the ones around today that many believe were reintroduced by park management. I have been told innumerous times, especially by older men, how the wolves before were bigger, stronger, hairier, fiercer than those "skinny ones the park releases". I have had people swearing that they saw a white van releasing wolves at night, despite the fact that wolf reintroduction has never occurred in Portugal. It is quite curious how this idea came to be such a strong conviction. According to Álvares et al., these rumours may be considered "as a means of cultural resistance, in which wolf adversaries perceive and actively challenge the dominant wolf recovery and protection paradigm" (Álvares et al. 2011, 323). It may indeed be interpreted as such, but what I want to emphasize is the difference established between the wolves from before and the ones from the present. If wolves *are* being reintroduced – or released as people say – they are not "as good" as the ones from the old days. This may indeed discredit the conservation programme as capable of delivering "good" wolves, but we must also consider that it can be related to a tendency to nostalgically romanticize the past. This belief also states that the wolves from before were worthy of admiration. Paradoxically, this may suggest that for some people, they have become less wild. Or maybe some people are not willing to let go of the idea of a ferocious creature: if they are not like that now, they certainly were in the past, so these have to be different ones. It may be a reaction to the dissemination of scientific knowledge that states that wolves do not attack people or kill animals for pleasure, as many believed in the past and some maintain today. Finally, this idea suggests that there is not a single straightforward traditional perception of the wolf immune to contemporary conceptions of nature and animal species.

The ecological wolf or managing the wild

Although biology affirms the capacity of the species to thrive in humanized landscapes, paradoxically, the wolf remains as a symbol of wilderness. Adjectives such as "mystic" or "magic" are often used in touristic discourses. What does this say about the concept of wilderness? In a time of rewilding movements, what do people mean when they yearn for wilderness? William Cronon (1996) is an incontestable reference when it comes to analysing this powerful modern myth. According to him, the idea of wilderness is an important romantic legacy that, more than a fact of nature, is a state of mind defined mostly by wonder. It is an idea that makes us project a fiction of untouchability and purity into the spaces and that persists upon the invented idea of wilderness experience, an experience independent of the real characteristics of the natural space. The idea that I find more frequently associated with the kind of nature in which people believe wolves can be found is precisely that of purity. To enjoy nature in "its purest state" or "pure nature" are common expressions among park visitors and are related to a state of non-human interference. Independently of the degree of humanization of the landscape or its rural character, the existence of wild species is considered a guarantee of a certain degree of untouchability. The presence of wolves, however, means human interference in terms of nature conservation. In many ways, conservation actually means restoration, since the objective is to manage ecosystems so they get as close to the original as possible; the original meaning, for instance, is free of alien species. Original according to Butler's notions of indigenousness, fixity and immutability, is an idea that perpetuates myths of "timeless natural utopias" (Olwig 2001, 342). Although "there is no other nature to get back to", as Kay Milton suggests (1996, 223), the concept of nature conservation held implicitly is that we can go back to or recreate a time of peaceful coexistence between humans and non-humans that existed somewhere in the past. To go back to a more natural state through landscape intervention (Harrison and O'Donnell 2010) or species reintroduction implies a selection and a categorization of appropriate elements and an idea or model of the type of nature we pretend to install.

Etienne Benson writes about the "managerial turn" (2016, 185) in nature conservation and states that "the endangerment sensibility in all of its varieties seems inextricably tied to a vision of species as frozen in time and of conservation as an effort to insulate them against change" (Benson 2016, 191). This "managerial turn" that happened in the 1980s is characterized by species management in their natural environment through GPS collars or veterinary exams or even through the reintroduction of animals born in captivity. Benson argues that this "can be seen as a reconciliation of wilderness preservation and wildlife management, but it also reflected a deeper shift in the understanding of nature" (Benson 2016, 189). In fact, Cronon states that "biological diversity (indeed, even wilderness itself) is likely to survive in the future only by the most vigilant and self-conscious management of the ecosystems that sustain it, the ideology of wilderness is potentially in direct conflict with the very thing it encourages us to protect" (Cronon 1996, 81, 82). This conservationist paradigm inevitably makes us question such categories as "wild", since this wilderness is maintained and managed by humans. Wilderness is a cultural, constructed concept and it can be actually recreated. Species, for instance, are selected and managed: in certain protected areas, they can be exterminated, reintroduced or protected so a predeterminate ecological situation can remain frozen in time. This demonstrates that ecological integrity is not scientifically given but socially and culturally negotiated (Beltran and Vaccaro 2010).

Let us consider two examples of how wild nature is actually allowed to be. Buller (2014) tells us about beavers in Scotland: while in Knapdale genetically approved and selected beavers from northern Europe were being reintroduced, beavers in Tayside, probably escaped from captivity, were occupying the territory in an unplanned and unmonitored fashion. The Tayside situation was not acceptable to Scottish Natural Heritage and there were concerns about inappropriate hybridization, among other things. Ironically the "(bio)secure beavers of Knapdale, in their designated and policed reintroduction space (…) for all their monitoring, predetermined biological suitability, and unequivocally wild location" were not reproducing as planned, but the "feral beavers of Tayside (…) less scientifically proper and actively sharing co-constituted space with the human occupants" were thriving (Buller 2014, 241). Another excellent example comes from the Catalan Pyrenees. Beltran and Vaccaro demonstrate how this territory became a laboratory where the concept of biodiversity is continuously reviewed and negotiated (2010, 174). Conservation programmes in the area support the reintroduction of Slovenian bears genetically identical to the extinct indigenous ones and the existence of the pandemic alpine marmots that expanded naturally from the Alps after 15,000 years of absence. At the same time, they defend the extermination of beavers that were illegally reintroduced by an environmental group, although beavers are endangered and considered endemic. The long-established yet invasive mouflon and the domestic goats that became wild are also considered a threat to ecological integrity. Although that had been their habitat for centuries, the goats are not considered part of the ecosystem.

For Cronon, to define an animal as wild is to place it in "the ultimate landscape of authenticity" (1996, 80). Authentic natural landscapes will provide a habitat for authentic wild animals. The case of the horses that live freely in the PGNP is quite interesting. They are an emblematic species in the park. Most of them are *Garranos*, an autochthonous breed. They are usually referred to as wild horses, although, to be exact, they are not wild but semi-domestic as they have owners (who actually receive money from governmental programmes for breed conservation). Most of the time, when I explained this to tourists, I could see how difficult and unpleasant it was for most of them to let go of the "wild" category. I see the same situation repeating over and over in my present fieldwork: visitors ask for the wild horses and expect to see them. Sometimes, the group guides explain they are not really considered wild, but most of the people do not look convinced and do not abandon the category. But if the *Garranos* might have an ambiguous categorization, the wolf is definitely in the wild category. The wildest one. In the news, they are frequently called "wild wolves" as if it was not redundant. For Henri Buller, even though wilderness is understood as a human construction, "there is still an inexorable sense of fixity in wild nature, whether it be in the repetitive vitality of natural growth and seasonal change, the indigenousness of wildlife, the seeming immutability of landforms" (Buller 2014, 238). The author argues that, as zoos, wild nature areas such as natural parks are spatial constructions where species are organized into categories that actually say more about humans than about the animals (Buller 2014, 234). Thus, to categorize an animal as wild has more to do with the way people think an animal should be than how it really is or the conditions it lives in.

These examples above demonstrate how categories that we take for granted, such as "wild", "domestic", "alien" or "autochthonous", can be reinvented and negotiated. Beltran and Vaccaro suggest that they imply a cultural re-elaboration of the concept of species (2010, 173). One can ask how long an alien species must remain until it can be considered autochthonous. And for how long does it have to be absent to be declared invasive. How far back should we go in time to recognize a species as indigenous? Bears have been extinct in Portugal since the seventeenth century. Would an eventual reintroduction programme consider them part of the ecosystem? What is the perfect nature model to follow and how long into the past must we go back to find it? These are extremely interesting questions that deserve more consideration that I can offer here. My intention is to illustrate how categories can be constructed and maintained according to different agendas. Bearing this in mind, I intend to stress how the concept of the wild maintains its strength in public opinion, even in extremely controlled natural conditions. Considering how the wolf is represented by environmental organizations, touristic companies and by PGNP institutional discourses is an interesting exercise to illustrate what kind of wildness is being communicated.

Following Frigolé (2010), I have presented this ecological wolf as an eco-symbol for biodiversity that in many ways opposes the traditional ambiguous wolf that it is supposed to substitute. It is the one that can be found featuring in conservation campaigns and touristic adverts, inserted in a frame of biological and ecological neutrality. In fact, analysing the wolf pictures that circulate among the public, either in environmental or touristic contexts, one notices that there is a tendency to portray this species in a manner that conveys a specific kind of wildness. A picture seems to be such a neutral device, a moment frozen in time, but it contains a huge amount of information. Instead of explaining what we normally see in those images, it is easier to elucidate about what we do not see. We do not generally see pictures of wolves looking aggressive, wolves sedated next to humans or in humanized landscapes, and we rarely see pictures of wolves wearing a GPS collar, although there are wolves being studied by biologists that carry them. Thus, wolves are presented mostly as "natural", meaning untouched by humans. They are also normally presented as not being a threat, as they do not look ferocious in the pictures selected to present them to the public. The images of the wolf circulating – either in printed material or on the web – are the result of a choice: they transmit certain characteristics and trigger certain emotions. A wolf with a GPS collar may seem less wild than one would like. In the same way, the picture of a pack devouring a buck may be wilder than is needed. The pictures chosen to lead people away from the vermin perception of the wolf have implicit a selection of information about the creature that guarantees the right image is communicated. Analysing these pictures, we realize they tend to project an image that is easy to feel empathy with. An image of a creature that is wild but not so wild that might be felt as a threat.

Environmental organizations stress the importance of education to defeat the evil vermin image. The selection of the right pictures serves that function, but the information that accompanies them is defined by the same choices: the scientific rhetoric found in educational or touristic texts is also the result of a selection that intends to portray a certain kind of creature. If information about their social life, such as co-operation feeding the cubs, tends to be emphasized, other facts, such as the expulsion of elements of the group, tend to be omitted. The way an animal is represented is always a result of choices which are not always deliberate. There are aesthetic concerns and moral ones behind any chosen picture or fact. In the case of the Iberian wolf in Portugal, I argue that the biological traits are culturally selected to present a creature that opposes the vermin image but is actually influenced by the noble one. This ecological wolf is not only a biological part of an ecosystem but a constructed and anthropomorphized creature with whom we tend to empathize.

The noble savage

The wolf is definitely one of those natural elements that were selected as heritage to take into the future. It was selected by nature lovers, public opinion, and actors making heritage. As a shared symbol that incorporates a set of ideas about wilderness, nature conservation and biodiversity, it fits the heritage category perfectly. It belongs to the realm of "endangered autochthonies" (Heatherington 2008, 13). The Iberian wolf is an authentic exemplar of national natural heritage. But this "wolf heritage", the one whose pictures are shown to children in school or in exhibitions about Portuguese biodiversity or even in the PGNP visitor centres, is not the ecological wolf or the traditional one. It is certainly not vermin. I argue that it is a combination of the first two. The wolf that is being transported into the future, or the perception of the wolf that is being cultivated, is that of the noble savage. It is wild, but it represents a type of wilderness that is kind to mankind. The noble savage represents the goodness in nature, the balance and the diversity it encompasses. Nevertheless, it maintains the mysterious aura of the traditional wolf. To some, it may still cause a shiver. The noble savage is a powerful symbolic image that shapes how people perceive and care for the Iberian wolf.

Even if we look into the past to select elements, the nature we choose to reproduce and the species we choose to preserve say a lot more about our future than our past. Thus, the wolf is simultaneously a species we decide to perpetuate in our landscapes and our imaginary. The concept of wilderness is also something people are not willing to relinquish. It is a very elastic concept that stands for the kind of nature that is appreciated and revered. To analyse these choices and the way people understand or represent creatures like the wolf or powerful concepts like wilderness offers us precious insights about the relationship people create with nature and also about the values that inspire nature conservation.

I mentioned that everyone has a wolf story. I do too. It may be the reason I am writing this today. Once, right at my door, on a very dark night, I saw a pack of wolves. It was extremely dark so I could only see their eyes and guess their silhouettes. There was only a fence between them and me. They were howling for quite some time. They were just a few metres away. It did something to me.

Bibliography

Álvares, F. 2006. Espécies emblemáticas e desenvolvimento rural: o potencial do lobo ibérico e da sua identidade na cultura popular. Paper presented at Jornadas de Debate sobre Biodiversidade e Mundo Rural: Perspectivas e Estratégias de Conservação da Fauna Selvagem organized by Associação ALDEIA/NEBUP, Oporto, April.

Álvares, F., J. Domingues, P. Sierra, and P. Primavera. 2011. "Cultural Dimension of Wolves in the Iberian Peninsula: Implications of Ethnozoology in Conservation Biology." *Innovation – The European Journal of Social Science Research* 24 (3), 313–331.

Beltran, O. and I. Vaccaro. 2010. "Un zoo en los pirineos. Paradojas de la patrimonialización de la naturaleza." In *Los lindes del patrimonio. Consumo y valores del pasado*, edited by C. del Mármol, J. Frigolé, and S. Narotzky, 169–189. Barcelona: Icaria.

Benson, E. 2016. "Endangered Birds and Epistemic Concerns: The California Condor." In *Endangerment, Biodiversity and Culture*, edited by F. Vidal and N. Dias, 175–194. London: Routledge.

Boitani, L. 2000. *Action Plan for the Conservation of Wolves (Canis lupus) in Europe*. Strasbourg: European Council.

Buller, H. 2014. "Reconfiguring Wild Spaces. The Porous Boundaries of Wild Animal Geographies." In *Routledge Handbook of Human–Animal Studies*, edited by G. Marvin and S. McHugh, 233–245. New York, NY and London: Routledge.

Coates, P. 2015. "Creatures Enshrined: Wild Animals as Bearers of Heritage." *Past and Present: Supplement 10 Heritage in the Modern World*, 273–298.

Cronon, W. 1996. "The Trouble with Wilderness." In *Uncommon Ground: Rethinking the Human Place in Nature*, edited by W. Cronon, 69–90. New York, NY: Norton.

Dressel, S., C. Sandstrom, and G. Ericsson. 2014. "A Meta-analysis of Studies on Attitudes Toward Bears and Wolves across Europe 1976–2012." *Conservation Biology* 29 (2), 565–574.

Frigolé, J. 2010. "Patrimonialization and the Mercantilization of the Authentic. Two Fundamental Strategies in a Tertiary Economy." In *Constructing Cultural and Natural Heritage. Parks, Museums and Rural Heritage*, edited by X. Roigé and J. Frigolé, 13–24. Girona: Institut Català de Reserca en Patrimoni Cultural.

Harrison, R. and D. O'Donnell. 2010. "Natural Heritage." In *Understanding Heritage in Practice*, edited by S. West, 88–126. Manchester: Manchester University Press.

Harrison, R. 2013. *Heritage. Critical Approaches*. New York, NY and London: Routledge.

Heatherington, T. 2008. "Cloning the Wild Mouflon." *Anthropology Today* 24 (1), 9–14.

Lopes-Fernandes, M., F. Soares, A. Frazão-Moreira, and A. I. Queiroz. 2016. "Living with the Beast: Wolves and Humans through Portuguese Literature." *Anthrozoös* 29 (1), 5–20.

Marvin, G. 2010. "Wolves in Sheep's and Other Clothing." In *Beastly Natures: Animals, Humans and the Study of History*, edited by D. Brantz, 59–78. Charlottesville, VI and London: University of Virginia Press.

Marvin, G. 2012. *Wolf*. London: Reaktion Books.

Milton, K. 1996. *Environmentalism and Cultural Theory: Exploring the Role of Anthropology in Environmental Discourse*. New York, NY and London: Routledge.

Olwig, K. 2001. "'Time out of Mind' – 'Mind out of Time': Custom Versus Tradition in Environmental Heritage Research and Interpretation." *International Journal of Heritage Studies* 7 (4), 339–354.

Sepkoski, D. 2016. "Extinction, Diversity, and Endangerment." In *Endangerment, Biodiversity and Culture*, edited by F. Vidal and N. Dias, 62–86. London: Routledge.

Vidal, F. and N. Dias. 2016. "Introduction. The Endangerment Sensibility." In *Endangerment, Biodiversity and Culture*, edited by F. Vidal and N. Dias, 1–38. London: Routledge.

6 Actualizing wolves

Environmental education settings as part of wolf management in Switzerland

Elisa Frank

Introduction

After more than 100 years of absence, wolves are back in Switzerland. Unknown and unfamiliar to most Swiss people and evoking fear and admiration simultaneously, the returning wolves challenge society. In the face of this situation, informing people about these newly arrived non-humans is seen as a key issue. That is where environmental education comes into play.[1]

In this chapter, I focus on four strategies of *actualizing* wolves in the context of environmental education. Wolves have been returning to Switzerland for about two decades and only since 2012 has reproduction and the establishing of wolf packs on Swiss soil been evolving. Therefore, it is not surprising that there are numerous moments of *actualizing* these newly arrived non-humans in environmental education settings, i.e. situations in which the actual being-here of wolves in Switzerland is emerging.

The data I present are based on 'thick inspections' in exhibitions in Swiss natural history museums and in zoos[2] and qualitative interviews with curators and taxidermists. In addition to these well-known institutions of environmental education, I also look at a third, younger and more unusual format: guided hiking tours through wolf territories, which I explored by doing participant observation and interviews with the providers.

Hereafter, I will examine these environmental education settings as exposing sites (Bal 1996, 1–12). "In expositions", cultural theorist Mieke Bal writes, "a 'first person', the exposer, tells a 'second person', the visitor, about a 'third person'" (Bal 1996, 3–4). According to Bal, exposing gestures (which she discovers in diverse settings and situations, not only in museums) connect two aspects: they "point to things and seem to say: 'Look!' – often implying: 'That's how it is'" (Bal 1996, 2). "The 'That's how it is' aspect", Bal continues, "involves the authority of the person who knows: epistemic authority" (Bal 1996, 2). Thus, exposure as a specific multimedial form of discursive behaviour (Bal 1996, 3) is always "an act of producing meaning" (Bal 1996, 2).

In her analysis of museum displays, Bal connects "the discursive strategies put into effect by the museum's expository agent" and the "meaning-making that these strategies suggest to the visitor" (Bal 1996, 7; cf. also Muttenthaler and Wonisch 2010, 78). Although the discursive strategies and the meaning-making can be distinguished, they are, nevertheless, closely linked to each other. It is this connection that I try to capture with the formulation *actualizing*: the actual being-here of wolves is emerging through various discursive – *actualizing* – strategies. By describing the discursive strategies of the environmental education sites, I want to point out, at the same time, the "actual processes of meaning production these strategies suggest to the visitors" (Muttenthaler and Wonisch 2010, 112; translated by the author). Meaning is not reflected by environmental education sites, but *produced* in and through practices of environmental education.[3] Following a relational approach focusing on practices, I thus, examine the moments of *actualizing* wolves in particular and environmental education sites in general as networks of humans, animals, objects and other materialities which produce knowledge about wolves through interaction(s) (Latour 1993; Niewöhner, Sørensen, and Beck 2012; regarding museums, e.g. Muttenthaler and Wonisch 2010, 111; von Bose et al. 2012, 11–12). I will argue that environmental education settings, as sites of knowledge production and entangled sites, are a significant part of wolf management in Switzerland.

Actualizing wolves by dissolving borders between the 'inside' and the 'outside'

The wolves coming back to Switzerland are not actually physically 'present' as living animals in natural history museums or zoos. Nonetheless, what the visitors are able to see in museums or zoos refers to these wolves outside the walls, respectively, the enclosure (Muttenthaler and Wonisch 2010, 77; Poehls 2012, 98; von Bose et al. 2012, 9, 11, 13). The border that exists with this 'outside' is dissolved in such settings in various ways to *actualize* the being-here of wolves in Switzerland.

Making the museum walls disappear[4]

The Natural History Museum of St. Gallen was reopened in a new building in November 2016. As in the old museum, one can see wolves in the new museum too, in a room entitled "In the bear's empire". A bear plays an important role in the legend of the foundation of the city of St. Gallen and is also the city's heraldic animal. As the museum director explained to me in the interview, he and his team decided to dedicate a separate room in the new museum to this animal so important to St. Gallen. In order to expand the subject of the bear, the museum made use of the bear's habitat – the forest – and complemented this room with other animals naturally sharing this habitat with the bear – among them two other large carnivores: the lynx and the wolf.

In this 'bear's empire', the forest itself is represented by vertically towering wood beams. Several mounted large carnivores are placed on and between these wood beams. But besides them, one can also spot objects referring to humans: a beehive, a bear-tested rubbish bin and a red bench (Figure 6.1). By these objects, a commonly used, shared space of humans and predators is displayed. The bear-tested rubbish bin and the beehive indicate that this living together is not always without conflicts, but that solutions can be found.

However, special attention must also be paid to the red bench, placed at the edge of the museum forest.[5] A howling wolf and some little wolf pups are placed behind this bench. The museum visitor who sits down on the bench has the forest with the mounted wolves and the other large carnivores at his/her back and looks outside through a very wide window stretching from the ceiling to the ground. On the outside, the visitor sees a meadow, a church, a couple of apartment buildings and some trees in the background. These three elements – the big window, the red bench and the visitor sitting on it – dissolve the border between the 'inside' and the 'outside' of the building. The arrangement creates a message regarding the wolves and the other animals in the museum forest that the sitting visitor knows to be behind him/her: they could – at any time – slip through to the other side of the window and enter the landscape outside the museum walls.

Referring to sociologist of space Martina Löw (2016), the arrangement of objects, visitors, mounted animals and the architecture of the building (big

Figure 6.1 "In the bear's empire", Natural History Museum of St. Gallen (29 November 2016).
© Elisa Frank

windows) can be understood as a *spacing* process which, in combination with the concomitant *synthesis* linking social goods and living entities, constitutes a new space: a hybrid space dissolving the border between the 'inside' and the 'outside'.[6] This is based on a processual and relational notion of space, according to which "the museum spaces re-form again and again, and are constituted by the interplay of (representational) practices, things, actors [...] and their readings of the museum" (von Bose et al. 2012, 11; translated by the author). Space, therefore, is no longer an "entity in which things happen", but it is *"that which happens"* (Rogoff 1997, 53, emphasis in original; translated by the author). Consequently, this concept of space allows one – in the example discussed here – to also analyse the visitor moving through the exhibition as a central part of the production of space and knowledge (Scholze 2004, 273–278; von Bose et al. 2012, 13) and, therefore, in dissolving the border between 'inside' and 'outside' and in constituting a new, hybrid space that *actualizes* the being-here of wolves.

Interactions between wolves in captivity and wolves living in the wild[7]

The Dählhölzli Zoo in the city of Bern has kept a pack of European wolves since 2011. On a tree-trunk just outside of the wolves' enclosure, the visitor finds an audio point, where s/he can listen to four wolves talking: the two leaders of the zoo-pack and two wolves living in the wild. The four wolves all bear a name in which the surname – matching the history of each individual – means 'wolf' in one of the four Swiss national languages: Amarouk Wolf, Juliette Loup, Luigi Lupo and Ladina Luf. The four wolves do not get into conversations with one another directly, but only interact indirectly:

> [Amarouk Wolf, Swiss German – Bernese dialect]: [...] I have been living in this new enclosure together with my elegant partner [Juliette] since 2012. I regularly use this tree here to mark our territory to make clear that this beautiful small piece of forest is already occupied. [He sniffs.] Gosh! Here is another wolf. [He sniffs again.] It is even cheeky enough to mark on my tree! Let's hope for its sake that it has already slipped away, if not it's going to have serious problems with me. [He pees.] Right, now it smells of me again, the leader of the territory. [He sniffs.] There's some-one else too. Oh, but I'm not worried about this one, that's a young wolf, no rival for me.
>
> (Inspection record Dählhölzli Zoo, 2016; translated by the author)

The four wolves, thus, come by the tree on which the audio point is fixed, one after another, and they detect who else has passed by here by sniffing. Wolves living in Switzerland in the wild pass by the tree too:

> [Luigi Lupo, German with an Italian accent]: Ciao, my name is Luigi Lupo and I come from Italy. When I was three years old, my father

chased me out of our territory. Since then, I've been looking for a new home. Unfortunately, this is quite difficult here in Switzerland. A lot of streets, railroads and settlements bar my way. [...] Besides, they don't seem to like me everywhere. I had found a beautiful place once, with a lot of sheep I could chase easily as they were not protected, but then someone shot at me! The bullet is still in my left hind leg. Hey, this looks like a very cosy place here! [He sniffs.] What a pity! It seems that someone is already living here. [He sniffs again.] Dear me! This is a clear marking of a powerful wolf. I'd rather not take up the fight with that one. But before I leave, haha [he laughs in an underhand manner], I will, nevertheless, leave my business card. [He pees.]

(Inspection record Dählhölzli Zoo, 2016; translated by the author)

The two zoo-wolves talking in the audio point are living individuals actually bearing these names (Amarouk and Juliette, the leaders of the wolf pack of the Dählhölzli Zoo), whereas Luigi Lupo is not an actual wolf who lives or had been living in the wild in Switzerland. He rather personifies a well-known type of Swiss wolf today: wolves immigrating from Italy on the search for their own territory. The same applies to the character Ladina Luf. She is the imaginary sister of the about 30 wolves who[8] have been born into the first Swiss wolf pack in the Calanda region. By alluding in the audio point to publicly well-known aspects of the lives of Swiss wolves living in the wild, the two wolf characters, Luigi and Ladina, are made plausible.

Based on this audio tree, it is suggested that wolves living in the wild and the leader wolves of the zoo-pack compete against each other for one and the same territory. The pack of the Bernese Dählhölzli Zoo, in this way, becomes a part of the Swiss wolf population; the border between the 'inside' (wolves in captivity) and the 'outside' (wolves living in the wild) becomes blurred and, thus, the wolves in the zoo – visible to the visitors – *actualize* their fellow species living in the wild for the public. Blurring the border between 'inside' and 'outside' in zoos may happen through architectural elements as well as stories told (Gisler 2016; Siegmundt 2015). In the case discussed here, it is the stories told in the audio tree that entangle and bridge two realities: the wolves in captivity and the wolves living in the wild.

Moving in the 'outside'[9]

A third environmental education format – guided hiking tours through wolf territories – tries not to let a border with the 'outside' emerge or rather to keep this border as porous as possible by taking place in the 'outside', i.e. in areas where wolves actually live. Such offers exist in the Calanda region in the cantons of St. Gallen and the Grisons, where in 2012, the first wolf pups in Switzerland were born since the extinction of the species at the end of the nineteenth century and the first of today's Swiss wolf packs established its territory.

Nevertheless, a certain border with the 'outside' also exists in this educational format. Although one is moving outside in an area well-known and proven to be a wolf territory, one will normally not see the wolves live, particularly as both of the wolf hiking tours that are offered at the Calanda and that I attended are explicitly designed to be no wolf safaris – even though the possibility of encountering wolves exists but happens very rarely. That is why there were moments on these hiking tours in which the 'outside', i.e. the wolves living in the area we were hiking through, was *actualized*.

In October 2016, I participated in a two-day "hiking tour in the Calanda wolves' habitat" (the official title), which was organised by the association CHWOLF.[10] The hike was guided by a retired gamekeeper of the region, G., who started to engage himself with wolves while still on duty and continues to inform people about wolves, collecting evidence and documenting events right up to today. During this hiking tour, the border with the 'outside' was dissolved again and again when we stopped at different places and G. told us about wolf tracks and other evidence of the wolves' presence he had found right here or about encounters with wolves he had had in this particular place.

The second wolf hike I attended (in July 2016) was organised by a small hiking tour provider and lasted three days. The hiking guide, D., and the wilderness instructor, A., who offered this tour had never had an encounter with wolves in the Calanda which they could have told us about, and they had very rarely found any evidence of wolves (tracks, excrement) in this area. Therefore, it was the two hosts with whom we stayed overnight or people we met on the tour who provided stories of wolf encounters for us.

Other moments in which the 'outside', i.e. the actual being-here of the wolves in this area, became concrete on both hiking tours occurred when the guides or some participants found wolf excrement while walking. Furthermore, on the hike with the retired gamekeeper, a participant found some bones which the gamekeeper identified as remains of a roe deer killed by wolves. These made clear to us participants that wolves were actually present in the area through which we were hiking and that they had stayed in exactly the same places where we were now, some hours or days ago.[11]

In addition, there was a moment in the hike with the private tour provider in which the actual being-here of wolves in the surroundings, i.e. the 'outside', suddenly became very tangible:

> On the Rossboden (P. 2040), A. [wilderness instructor] spots a wolf-like animal on the opposite slope, near a little cabin (P. 1993), just above the tree line (linear distance ca. 600 m). All of us stop immediately, we look through our binoculars and start racking our brains. The animal is clearly dog-like, and the colour of its fur and its size seem wolf-like to all of us. But its behaviour confuses us: the animal is walking back and forth, in between times, it sits down and looks around, and all of this in the immediate vicinity of the little cabin, next to which a car is parked. Finally, when after about five minutes

people come out of the cabin and the animal jumps over to them and they stroke it, it becomes definitely clear that it is a wolfdog or another dog looking very wolf-like at such a distance.

> (Participant observation record hiking tour A. and D. 2016;
> translated by the author)

The moment just after A. had spotted the animal and we all started to observe and puzzle over it, placed us quite unexpectedly in the middle of the 'outside' and *actualized* the being-here of wolves.

Even in an educational format which takes place in the 'outside', there is still a border accompanying this 'outside' that needs to be overcome. The area hiked through actively needs to be established as a wolf territory by a correspondent performance. Compared to museums and zoos, different possibilities and moments exist or can be created in such a setting to dissolve this border between 'inside' and 'outside', as in these settings other social goods and living entities (e.g. wolf excrement, tracks, bones, persons having found evidence of the wolves' presence or having seen wolves, supposed wolf) are available for *spacing* and *synthesis* (Löw 2016). The Calanda wolf territory does not constitute the "background or platform of action", but, following a processual and relational notion of space (which conceptually integrates space in the course of action, cf. Löw 2016, 229) and referring to an understanding of walking (de Certeau 1988, 91–110; Tschofen 2013) and guided walks (Kowaleski Wallace 2006, 43–65; Korte and Paletschek 2009, 43–44) as an epistemic practice, it comes into being through the performance, i.e. the practices and actions of the tour guides and the participants.[12]

Actualizing wolves by involving people

Imparting to recipients of environmental education that the return of the wolves has very concretely to do with their own lives is a second type of *actualizing* the wolves' presence in Switzerland. This strategy deals with the changes the returning wolves cause potentially for each of us. Thus, the recipients are not only addressed, but they come into being as human beings concerned by the return of the wolves by learning specific practices they should perform as involved beings.[13]

Protection of herds makes every hiker an involved being[14]

At the end of the three-day hiking tour with the private tour provider, we met a sheep farmer who works with livestock guardian dogs. The encounter took place on the pasture where the sheep and dogs were staying. Thus, we could experience the livestock guardian dogs live in action: how they bark at you, examine you, come near you, assess you. In addition to this 'live performance', we had the possibility to ask the sheep farmer all our questions about sheep, wolves and livestock guardian dogs.

In the interview we carried out some weeks after the tour, D., the hiking guide, explained why this element of the wolf hike is so important to him:

> D.: We too are a part of this nature. We belong to it, we live in the nature, of the nature, so we should also have instruments to deal with this nature. The wolf is a part of it, we are a part of it. [...] We have the right to be a part of it and to not be excluded from it. [...] It's not about saying, the wolf is back, so farmer move away, buzz off with your sheep. But it's about saying, well, farmer, or well, society, the wolf is coming, it belongs to us as a part of nature; let's learn how to deal with it with the help of the methods we know. And the farmer, who keeps sheep, who keeps goats, his method is a shepherd or livestock guardian dogs or fences. And if we, our society, wants to have them [wolves], then we need to grant to the farmer that he can use these methods. The solution is not to say, wolf come back, and farmer keep your sheep somehow, [...] but I don't want livestock guardian dogs barking at me when I'm hiking.
>
> (Interview A. and D. 2016; translated by the author)

The participants of the tour – who are often enthusiastic hikers and partly joined this wolf hike primarily because of the hiking not because of the subject (wolves)[15] – will notice that the return of the wolves has something to do with their own lives. While practising their hobby, hiking, they may encounter livestock guardian dogs whose task is to protect sheep herds. The participants of the tour come into being at this moment as persons involved in the return of the wolves.

The aim of the encounter with the livestock guardian dogs during the guided wolf hike is not only to let the participants realise that they are a part of the process 'return of the wolves', but also to teach them how they can and should cope with this involvedness to contribute to a successful return of the wolves:

> A.: We want to impart an additional value, that is that people coming back from our tour know, aha, this dog is big, it's frightening, it barks, but I, I have my methods too, I cannot silence it, but when I'm behaving in an adequate manner, then nothing will happen to me. [...] It's important to us to teach this to people who... or rather to anyone who spends time outside, in nature. [...] Maybe I'm talking very much as a wilderness instructor now for whom it's a lot about bringing humans in touch with nature and leading them to the idea that they are a part of it. [...] That is to say, I don't have to run away, I don't have to accept to be excluded, but I too can feel at home here. And to know how to face a livestock guardian dog, of course, helps me to feel at home when hiking.
>
> (Interview A. and D. 2016; translated by the author)

Thus, after having met the sheep farmer and his livestock guardian dogs, participants should know how to behave in front of these dogs, so that no incidents will occur, and they will be able to continue practising their leisure time activities.

This example makes it very clear that environmental education is not only about imparting knowledge but also about producing knowledge and generating corresponding practices. At the moment of interaction with the animals (sheep, livestock guardian dogs) and other people (e.g. the farmer), the participant comes into being as a particular human being: a person involved in the return of the wolves who has a task to perform and, accordingly, needs to adapt some of his/her practices. This knowledge is literally incorporated in the case of the hiker who encounters livestock guardian dogs when this person tries to keep his/her body calm facing such dogs. Possibly some of the person's spatial movements also change; that is the case when a person makes a detour to avoid encountering a protected sheep herd, for instance, because s/he is frightened of dogs or s/he has a companion dog which complicates encounters with livestock guardian dogs.

The statements of D. and A. show that a particular concept of nature underlies this 'involving people': together with non-human actors (e.g. wolves, game, livestock guardian dogs, sheep, plants), human beings should perceive themselves as equal components of nature, provided with rights and duties. Thus, being involved is related to tasks, but it especially also means inclusion. If the participants of the wolf hike put into practice what is intended by the providers, then they will perform nature as perceived by the hiking guide and the wilderness instructor, and they will, thus, let this nature come into being. At that moment, this environmental education offer not only imparts knowledge about wolves and nature to the recipients but is clearly part of the production of (a particular) nature.

A new waste behaviour for all of us[16]

Something similar – even though it is more to do with bears – happens in the temporary exhibition "Living with Big Predators" (May to November 2016) in the Museum of Nature in Olten. The exhibition consists of a lot of text panels with graphics and photographs, staged preserved specimens and a few objects. One of these objects is a bear-tested rubbish bin.[17] The visitor is allowed to touch and use this object. The corresponding 'operating instructions' are an integral part of the original object itself as they are attached to the bin: "Pull and turn left around". Due to the four languages in which this sentence is stated – German, Italian, Rhaeto-Romanic and English – the sticker also reveals to the visitor that this object must come from the Grisons.[18]

This bear-tested rubbish bin is a sort of 'simultaneous' object displayed in a museum in a small town on the Swiss Plateau (Olten) and, at the same time, one could come across it in a valley in the Grisons where bears may be present. In interaction with this object, the museum visitors experience what may be their personal contribution to the return of the large carnivores: to dispose of their rubbish correctly when crossing bear territory, which the visitors can situate in Switzerland, or rather in the Grisons thanks to the sticker. Thus, the museum visitors experience themselves in this interaction as human beings involved in the return of the large carnivores and learn the corresponding practices, i.e. they learn what the object they should look for is like and get to know how to handle it. The knowledge imparted at the museum will influence the practices of the museum visitor in bear territory in a very concrete and direct way. The exhibition can, thus, be understood as a site of large carnivore management.

Actualizing wolves by regionalizing

Furthermore, *actualizing* wolves in the context of environmental education takes place by locating wolves – on the one hand, in specific, geographically locatable regions and, on the other hand, in conceptional landscapes. Firstly, I will look at regionalizing as a form of *actualizing* wolves that intends to increase the localness – and along with it, the actual being-here – of wolves.

Waiting for a local wolf in St. Gallen[19]

At the beginning of our conversation, the director of the Natural History Museum of St. Gallen explained to me his understanding of the institution: "We are a regional museum. We are the archive of nature of the canton of St. Gallen" (Interview Director Natural History Museum St. Gallen 2017; translated by the author).[20] Corresponding to this perception that the building is hosting an archive of the nature of the region, the newly reopened museum shows the local fauna in its central room "From the Lake of Constance to the Ringelspitz".[21] The centre of the room is occupied by a big plastic relief model of the canton, and various mounted animals are displayed along the walls, ordered according to different local habitats.

As has already been mentioned, in 2012 the first wolf pups were born in Switzerland since the extinction of the species at the end of the nineteenth century. The territory of this first Swiss wolf pack extends over the cantons of St. Gallen and the Grisons. But a wolf is not to be seen in this panorama of the endemic fauna in the natural history museum of St. Gallen – not *yet*. During the interview, the museum director showed me the place which is saved especially for the wolf in this regional panorama: between ibex, mountain hare and Alpine marmot, the wolf will one day find its place here in the habitat entitled "Mountains and Peaks".

But why is the museum waiting to exhibit a wolf here? The reason is that they want to display here a local, 'native' wolf who lived in the region in the wild, thus, a 'proper' wolf of St. Gallen: "That's arranged with the head of the [cantonal hunting] authorities. If a wolf is shot or if an animal dies who may still be mounted later, it will have its little place here" (Interview Director Natural History Museum St. Gallen 2017; translated by the author), the museum director explains. So, the museum is waiting for a wolf to perish on cantonal soil one day whose body is in a condition that still allows taxidermy of the animal. This must not be misinterpreted as the museum really waiting and actually hoping for a wolf to be killed in their canton. But the museum thinks it is realistic that it will happen sooner or later (as licences for shooting wolves can be issued for several, legally regulated reasons in Switzerland; cf. BAFU 2016, 11–12) and took that in consideration when planning the new exhibition.

The 'localness' of wolves is increased even more by the museum's intention to display a wolf of 'proper' St. Gallen origin and, therefore, to save the necessary square metre (for which the wolves returning to the canton of St. Gallen are just as responsible as the museum people);[22] wolves come into being as a genuine part of the canton of St. Gallen. The space emerging here – co-produced by humans and wolves – and the hybrid object that goes with it (a mounted wolf of proper St. Gallen origin) show how the museum and its collection are entangled with wolf management; the museum and its taxidermy workshop become a direct link in a regional (dead) wolf management chain. By identifying the spatial tracks of a collection beyond the built storage place, the museum depot's boundaries are expanded and its entanglement with other spaces and sites outside the museum walls can be traced (Schneider 2012).[23] That is how it becomes possible to understand museums as part of a regional wolf management network.

M44: the Wolf from Domleschg[24]

The regionalization of wolves, as observed in the natural history museum of St. Gallen, also takes place in another case – the one of wolf M44,[25] whose regionalization is closely connected to an individualization.

M44 was shot accidentally by a hunter in the winter of 2014 in the Domleschg, a valley in the Alpine canton of Grisons. After his death, the pathological, ballistic and genetic examinations usually carried out in such cases took place. A few days after the accidental death, the President of the Foundation Council of a Domleschgian museum deposited an application that his museum would get this wolf's body. The application was approved by the cantonal government. The contract to mount the wolf was awarded to Sabrina Beutler, a freelance taxidermist who is now processing the order.

During our first meeting, the taxidermist told me in greater detail about the story of M44 and his life after death outlined above. How the taxidermist talks about this wolf is very significant for her perception and dealings with the animal entrusted to her. She has a specific understanding of this wolf due to his story. M44 is, Beutler explains in an interview,

> an individual, not just any old wolf, for it is M44, *the* wolf from Domleschg. The sudden appearance of this wolf caused a considerable stir. Taxidermy ultimately makes it possible for the public to meet this animal themselves. They can stand in front of it, look it straight in the eyes and engage personally with the animal.
> (Alpines Museum der Schweiz and Universität Zürich – ISEK 2017, 28; translated by Pauline Cumbers)[26]

Understanding M44 as an individual and a regional Domleschgian wolf influences the taxidermist's practices very concretely:

> In the case of M44, the technique used matches the significance of an individual that will never come to life again and is indeed irreplaceable. Thus, the animal's skin is tanned separately and not treated in a mass process – thrown together with some sheepskins, for example. The artificial corpus that provides the internal core of the preserved specimen is built by me from natural products that are durable. That is to say, I do not just use any artificial substance to hand without knowing if it will simply fall apart in 30 years' time.
> (Alpines Museum der Schweiz and Universität Zürich – ISEK 2017, 29;
> translated by Pauline Cumbers)

On the one hand, the regionalization in the case of M44 happens – as is also the case for the St. Gallen example – through the practices of collecting and displaying of a local museum. However, in the case of M44, the regionalization is, on the other hand, also realised through specific preserving techniques, of which the result – a long-lasting preserved wolf – also contributes to a 'Domleschgisation' of wolves in general.

The knowledge about M44 as an individual and a Domleschgian wolf and the taxidermy practices interact very closely with each other as the taxidermist explains:

> The attention he [M44] receives and his importance in the mind of the public at the present time raise him to the status of a cultural asset. In these cases, I have more responsibility and have to guarantee that the preserved specimen will last for hundreds of years and will still make people aware of M44 in two hundred years' time.
> (Alpines Museum der Schweiz and Universität Zürich – ISEK 2017, 29;
> translated by Pauline Cumbers)

In Beutler's eyes, due to his story, M44 is a Domleschgian cultural asset and, conversely, she materializes this perception through her work and, by doing so, guarantees that M44 will stay a cultural asset for the next generations.[27] M44 in particular and wolves in general are and become regional – knowledge is practice (Niewöhner, Sørensen, and Beck 2012, 40). The taxidermy workshop becomes, in this respect, a central part in the museum's production of meaning.[28]

Actualizing wolves by showing their place in a conceptional landscape

Not only is locating wolves in geographically identified regions a technique used to *actualize* the presence of wolves in the framework of environmental education offers, but so is showing their place in a 'conceptional' space. This is closely connected to an important discussion in the context of the Swiss wolf debate: is there any place at all for wolves in densely populated and overly settled Switzerland? Where would it be? Do wolves find their way around our cultural landscape (*Kulturlandschaft*)? Are wolves threatening this cultural landscape by their presence? Perceiving and thinking about the environment in conceptional spatial categories such as 'nature' and 'culture', 'wild' and 'domesticated' is characteristic for modern Western thought (Descola 2013) and in these spaces, too, wolves are located – and in this way *actualized* – in the context of Swiss environmental education sites.

The wolf in the visual cultural landscape[29]

The Natural History Museum of Fribourg showed a temporary exhibition entitled "Wolf – Back among Us" from September 2016 to August 2017. The title corresponds with the visual elements of the exhibition's poster (Figure 6.2): the silhouette of a wolf filled with the photograph of a landscape. The museum director and curator of the exhibition told me that he paid attention to the type of landscape in the poster: "We decided that in the silhouette of the wolf, there should be a landscape in which it [the wolf] could definitely live. And that is the region Schwarzsee, which is recognisable for some of the Fribourg people in this photograph. And then again, here a chair lift" (Interview Director Natural History Museum Fribourg 2016; translated by the author). The wolf is, thus, placed in this poster not only in a specific regional landscape (Schwarzsee – a region in the canton of Fribourg), but also in a conceptional landscape: a cultural landscape (*Kulturlandschaft*) co-shaped and co-used by men (chair lift, furthermore, a house and a small street are to be seen on the poster, cf. Figure 6.2).

Regarding visual elements in the exhibition, the curators paid attention to them, always showing a landscape where traces of men are clearly recognisable, for instance, in the big relief-like illustrations. The curators commented on the first sketches the illustrator presented as follows:

MUSEUM DIRECTOR: We explained, for example, that we wanted to have some high-rise buildings in the background, and that we wanted to see a high-tension line, because we wanted to break out of this idea of, of…
EF: Of the wolf out in the wild.
MUSEUM DIRECTOR: Exactly.
(Interview Director Natural History Museum Fribourg 2016;
translated by the author)

The aim of these visual markers was to show the visitors that the wolf "is capable of living in a landscape which is no longer nature or which no longer corresponds to this unspoiled nature. That's an idea we absolutely want to correct" (Interview Director Natural History Museum Fribourg 2016; translated by the author). The wolf is deliberately placed visually in a particular conceptional space. The being-here of wolves is *actualized* by showing the

Figure 6.2 "Wolf – Back among Us", exhibition poster, Natural History Museum of Fribourg.
© Musée d'histoire naturelle Fribourg

wolves' place in the Swiss landscape, as we understand it, through conceptional spatial categories. The recipients learn to perceive this landscape as a
space shared by humans and wolves.

Wolves in the human area, humans in the wolf area[30]

A similar and yet more complex idea of the wolves' place in our landscape was
imparted on the two-day hiking tour guided by the retired gamekeeper mentioned already. This hiking tour started in Tamins, one of the villages at the foot
of the Calanda massif, and brought us to the Kunkelspass, where we stayed
overnight. The next day, we continued our ascent from the Tamina valley up to
the Ringelspitz mountain hut. From the hut, we descended the so-called Lavoi
valley back to Tamins. Shortly after having started at the bus station in Tamins,
we stopped, so that our guide, the retired gamekeeper, could show us something:

> At the edge of the village, before leaving Tamins, G. [retired gamekeeper]
> points out to us – pointing at the landscape surrounding us [Figure 6.3] –
> the specific structure of the villages around the Calanda massif. Appar-
> ently, G. states, the residential areas here reach the edge of the forest. In

Figure 6.3 Stop at the edge of the village of Tamins, guided hiking tour in the Calanda
wolf territory (8 October 2016).
© Elisa Frank

this respect, G. explains, it is not surprising and is normal that the wolves may be seen from time to time close to the villages, especially because wolves like to use human infrastructures, such as streets or bridges, as the latter also make their moving easier. In addition, during wintertime, wolves follow their prey moving down to the valley floor, and following the prey, the wolves choose the direct way, possibly leading through and not around the village. G. assures us that this is normal and not problematic wolf behaviour.

(Participant observation record hiking tour CHWOLF 2016; translated by the author)

This picture of wolves moving naturally through human cultural landscapes (*Kulturlandschaften*) and residential areas, is contrasted by a scene on the second hiking day:

Shortly before reaching the top of our hike, the Ringelspitz mountain hut, we stop again and G., pointing down to and out of the Tamina valley we came from, comments on this – partly cloudy – landscape [Figure 6.4]: If one sees that, it is obvious after all that there is enough space for wolves in Switzerland.

Figure 6.4 Stop with view over the Tamina valley, guided hiking tour in the Calanda wolf territory (9 October 2016).
© Elisa Frank

And G. announces that we will see more of it on our descent through the Lavoi valley in the afternoon [...] During one of our stops in the Lavoi valley [Figure 6.5], G. explains why the wolves like to stay here: It's steep, there's plenty of game around and it offers a lot of retreat areas. Humans, G. continues, come to the valley, but very canalized on the hiking trail, as the terrain is so steep that we humans scarcely move voluntarily off the beaten track. Furthermore, these human activities in the Lavoi valley – hiking, seldom biking – only take place during the day and in particular seasons.

> (Participant observation record hiking tour CHWOLF 2016;
> translated by the author)

At first sight, these two scenes seem to contradict each other. Rather than a contradiction, this is a clue for a specific perception of space in a conceptional sense and its use by human beings and wolves. In the eyes of the retired gamekeeper, something like a wolf area and a human area do exist, but the wolf area is not reserved exclusively for wolves, just as conversely, the human area is not reserved exclusively for human beings. However, it is predictable for both sides when one enters the other's 'own' area. There are reasons for humans to move into the wolf area, for instance, for hiking, biking,

Figure 6.5 Stop in the Lavoi valley, guided hiking tour in the Calanda wolf territory
(9 October 2016).
© Elisa Frank

cultivating alps or harvesting timber, but this normally happens during particular periods of days and seasons, and generally also on predictable tracks and always in the same 'mini-areas', for example, a hiking trail or a mountain pasture. Vice versa, wolves appear in the human area from time to time, due to specific reasons and activities, on predictable tracks and at predictable times, for instance, when they follow their prey down to the valley floor in winter and, in doing so, use human infrastructures.

The retired gamekeeper reads and imparts the landscape we are moving through in the situations described above as a conceptional landscape. By locating wolves and human beings in this conceptionally perceived space and by assigning to each of them – depending on activities, daytime and season – their places in it, he *actualizes* – in interaction with the surrounding landscapes – the being-here of the wolves to the participants. The latter learn to read the landscape they hike through as a shared space in which a living together of wolves and humans is possible – provided that both of them stick to certain rules when entering areas primarily inhabited by the other who, vice versa, tolerates the former. This is based on a specific, relativizing concept of the boundary between wilderness and cultural landscape (*Kulturlandschaft*) according to which such a boundary exists but is permeable (Frank and Heinzer 2019). Similar to the case of the hikers involved, when the participants of the wolf hike start perceiving the landscape as imparted by the tour guide and acting accordingly, a particular nature comes into being.

Conclusion

Concluding my article, I would like to discuss in what way environmental education offers – by *actualizing* wolves in the four ways presented above – become themselves a part of wolf management in Switzerland. There are two aspects I would like to point out.

Firstly, looking at environmental education settings as networks of humans, animals, objects and other materialities emphasises that these settings are not just 'simple' sites where 'factual knowledge' about wolves is imparted, but they are also multilayered, complex sites where knowledge about wolves is actually produced by humans, animals, objects and other materialities through interactions.[31] Museums, zoos and guided hiking tours through wolf territories generate stocks of knowledge that have the potential to influence practices, since, at the moments of interaction, different particular wolves, humans and natures come into being. The recipients of environmental education offers are not only addressed but constructed as involved human beings and the knowledge imparted concretely influences their practices. By encountering livestock guardian dogs or interacting with a bear-tested rubbish bin, the recipients experience themselves as being part of the return of the large carnivores, and they learn specific practices to make their own contributions to this return (*involving people*). When wolves are displayed in museums in local cultural landscapes (*Kulturlandschaften*) and participants of wolf hikes experience the area they are walking through as spaces commonly used by wolves and humans following predictable patterns, then human beings and wolves come into being as neighbours who respect and tolerate each other, and yet stick to certain rules to be predictable for one another (*showing the wolves' place in a conceptional landscape*).

Secondly, looking at environmental education settings as networks of humans, animals, objects and other materialities sharpens the understanding of how much these are entangled sites (Latour 1993; Kirksey and Helmreich 2010). When mounted wolves, objects, visitors and the architecture are arranged in such a way that the museum walls disappear, when wolves in captivity interact with wolves living in the wild, and when the 'real' wolves occur in different respects on guided hiking tours through wolf territories, then the border between the 'inside' and the 'outside' becomes blurred and a hybrid space is created for which boxes labelled 'natural' or 'cultural' seem to make little sense (*dissolving borders*). Furthermore, museums participate in the collective dealings with wolves and become a direct link in wolf management chains when they offer dead wolves who lived in their surroundings a life after death by converting these wolves' bodies into 'native' taxidermy specimens – already extremely hybrid objects as such – for the regional public (*regionalize*).

These two points make clear that environmental education settings – as sites of knowledge production and entangled networks – are not sites detached from the reality of wolf management in a narrow sense, but that they are, on the contrary, a significant part of wolf management in Switzerland. The examples outlined in this article show how environmental education settings interrelate with other fields of society.[32] It is not only on mountain pastures, in public administration or in the forest that the return of the wolves is managed, but also in natural history museums, collections, taxidermy workshops, zoos and on guided hiking tours through wolf territories.

Notes

1 In the framework of the research project "Wolves: Knowledge and Practice. Ethnographies on the Return of Wolves in Switzerland" (project leader: Bernhard Tschofen; project staff: Nikolaus Heinzer and Elisa Frank; project number: 162469), funded by the Swiss National Science Foundation, I examine, among others, environmental education offers regarding wolves. I thank all my field partners for letting me participate in their thoughts and practices. Thanks to Michaela Fenske, Bernhard Tschofen and Nikolaus Heinzer for helpful comments on the text, and to Philip Saunders for the proofreading.

2 The notion 'thick inspection' – following Clifford Geertz' 'thick description' – stems from Smilla Ebeling (2017). I borrow the term, but I do not conduct my thick inspections in natural history museums and zoos exactly as Ebeling did in her fieldwork in local museums. Furthermore, my fieldwork in natural history museums and zoos is inspired by the works of Mieke Bal (1996, esp. 13–56), as well as Roswitha Muttenthaler and Regina Wonisch (2010).

3 For museums as meaning producing sites, cf., e.g. Muttenthaler and Wonisch (2010, 78, 82); von Bose et al. (2012, 9).

4 This section is based on the following data: Inspection record Natural History Museum St. Gallen 2016, Interview Director Natural History Museum St. Gallen 2017.

5 von Bose et al. (2012, 15) also point out that departing from seating furniture may be a revealing way of describing and analysing exhibitions.

6 The fact that the bench the visitors may sit on is a so-called '*Landi*-bench' reinforces this message. The *Landi*-bench was designed for the Swiss national exhibition in 1939 (called '*Landi*' as an abbreviation for '*Landesausstellung*'), cf. BURRI (undated). Since then, this bench has been widespread in cities, on fields, in parks or in the mountains all over Switzerland and, therefore, should be a well-known element of Swiss landscapes for most of the museum visitors.

7 This section is based on the following data: Inspection record Dählhölzli Zoo 2016.

8 I use gendered pronouns in cases where I know the sex of the individual animal as it was mentioned by my field partners. In all other cases, I use the impersonal pronoun "it" – being aware of its problematic aspect of reducing the animal to a passive, mechanical status. Always choosing the relative pronoun "who" (instead of "that") is a way not to increase this further (on this difficulty, cf. Fudge 2017, 268–269).

9 This section is based on the following data: Participant observation record hiking tour A. and D. 2016, Interview A. and D. 2016, Participant observation record hiking tour CHWOLF 2016.

10 CHWOLF is an association supporting the returning wolves in Switzerland in various ways, cf. CHWOLF (undated).

11 It was not possible for the hiking guides to plan such finds, but the probability of discovering evidence of wolves, especially excrement, is fairly high in the Calanda area.

12 Furthermore, theoretical considerations on authenticity (and tourism) (for an overview cf. Schäfer 2015, 15–44) can serve as analytical tools to understand the process of making the Calanda into Calanda wolf territory. In my doctoral thesis, I also make use of such concepts to analyse the space constituting process occurring on guided walks.

13 For observations on different types of *involvedness* in the context of the return of the wolves in Switzerland, cf. the chapter of Nikolaus Heinzer in this volume as well as Frank, Heinzer, and Tschofen (2016).

14 This section is based on the following data: Participant observation record hiking tour A. and D. 2016, Interview A. and D. 2016.

15 We learned about that on the wolf hike we attended ourselves in 2016 and the providers of the tour confirmed this in the interview for the other years too and for theme walks they provide in general.

16 This section is based on the following data: Inspection record Museum of Nature Olten 2016.

17 As mentioned in the section 'Making the museum walls disappear', a bear-tested rubbish bin is also displayed in the room 'In the bear's empire' in the Natural History Museum of St. Gallen. The following considerations are, thus, also valid for the situation in St. Gallen.

18 Rhaeto-Romanic, one of the four Swiss national languages, is only spoken in the Grisons.

19 This section is based on the following data: Inspection record Natural History Museum St. Gallen 2016, Interview Director Natural History Museum St. Gallen 2017.

20 St. Gallen is a city in the east of Switzerland, but it is also the name of a canton, so, St. Gallen also stands for the region surrounding the city of St. Gallen. The museum is in the city.

21 The Ringelspitz (3247 m) is the highest summit in the canton of St. Gallen.

22 I look in detail at wolf taxidermy taking place in the context of the return of wolves in Switzerland as a multispecies process in another paper (Frank 2020), asking whether there are moments of wolfish agency in the unquestionably very human-dominated process of taxidermy.

23 Object biographies of taxidermy specimens which show how these objects "accrue meaning as they move between and interact within various social contexts" (Poliquin 2008, 129) can be a productive method to detect such entanglements.
24 This section is based on the following data: Field notes meetings with S. Beutler in her workshop and in Zurich 2016, Field notes phone call President of the Foundation Council of the Domleschgian museum 2016, Alpines Museum der Schweiz and Universität Zürich – ISEK (2017, 28–29).
25 M44 is the 44th male wolf identified in Switzerland by DNA analysis since the return of the species.
26 The Swiss Alpine Museum in Bern showed the exhibition "Der Wolf ist da. Eine Menschenausstellung" ("The Wolf Is Here. An Exhibition about People") from May 13 to October 1, 2017. The exhibition is a co-production of the Swiss Alpine Museum and our research project "Wolves: Knowledge and Practice" (cf. note 1). A central element of the exhibition are eight audio points where different experts whose jobs bring them into contact with wolves talk about their experiences, among them the taxidermist Sabrina Beutler. The audio points were produced by Michael T. Ganz after an intensive briefing by Nikolaus Heinzer and me, transcribed by Elena Lynch and translated into English by Pauline Cumbers. The transcripts in German can be read up on Alpines Museum der Schweiz and Universität Zürich – ISEK (2017). I make use of these concise quotes of the audio point here as my conversations with Beutler up to now have not been recorded on tape but documented in the form of field notes and records taken from memory.
27 On the subject of taxidermy specimens as historical and cultural records, cf. the contributions in Alberti (2011).
28 von Bose et al. (2012, 15) also point out that workshops are significant museum spaces.
29 This section is based on the following data: Inspection record Natural History Museum Fribourg 2016, Interview Director Natural History Museum Fribourg 2016.
30 This section is based on the following data: Participant observation record hiking tour CHWOLF 2016.
31 Cf. Niewöhner, Sørensen, and Beck (2012, 20), who stress that such "areas of intersection are not to be understood as spheres in which a simple translation or popularisation of scientific knowledge for a 'lay public' takes place [...]. Rather these areas of intersection are conceived as zones of 'transaction'" (translated by the author). For exhibitions and museums of science as contestable sites in which knowledge and truth are produced and defined and which, therefore, should be examined by looking in detail at practices, techniques and tactics cf. Macdonald (2002; 2006). In her ethnography of the making of an exhibition in a science museum, Sharon Macdonald detects that the communication model she had in mind at the beginning of her work – "a model in which science was taken from the world of science and translated by the museum into something to be 'responded to' by the public" – turned out to be "far too neat in practice" (Macdonald 2002, 6).
32 This mechanism has been repeatedly stated for museums, cf., e.g. von Bose et al. (2012, 13); Poehls (2012, 91).

Bibliography

Alberti, S. J. M. M., ed. 2011. *The Afterlives of Animals. A Museum Menagerie.* Charlottesville, VA: University of Virginia Press.
Alpines Museum der Schweiz (Hächler, B.) and Universität Zürich – ISEK (Tschofen, B.), eds. 2017. *Der Wolf ist da. Eine Menschenausstellung.* Bern: Alpines Museum der Schweiz.

BAFU: Bundesamt für Umwelt, 2016. "Konzept Wolf Schweiz. Vollzugshilfe des BAFU zum Wolfsmangement in der Schweiz." https://www.bafu.admin.ch/bafu/de/home/themen/biodiversitaet/publikationen-studien/publikationen/konzept-wolf-schweiz.html [accessed 28 May 2018].

Bal, M. 1996. *Double Exposures. The Subject of Cultural Analysis*. New York, NY: Routledge.

BURRI public elements AG, undated. "Landi." http://www.burri.world/de/mobiliar/landi/sitzbank-mit-rueckenlehne [accessed 23 November 2017].

CHWOLF, undated. "Der Verein CHWOLF." https://chwolf.org/ueber-uns/verein-chwolf [accessed 23 November 2017].

de Certeau, M. 1988. *The Practice of Everyday Life*. Berkeley, CA: University of California Press.

Descola, P. 2013. *Beyond Nature and Culture*. Chicago, IL: University of Chicago Press.

Ebeling, S. 2017. *Durch die Blume. Geschlechternarrationen in musealen Naturdarstellungen*. Münster: Waxmann.

Frank, E. 2020. "Multispecies Interferences. Taxidermy and the Return of Wolves." *Ethnologia Europaea* 49 (2).

Frank, E. and N. Heinzer. 2019. "Wölfische Unterwanderungen von Natur und Kultur: Ordnungen und Räume neu verhandelt." In *Ordnungen in Alltag & Gesellschaft. Empirisch-kulturwissenschaftliche Perspektiven*, edited by S. Groth and L. Mülli, 93–124. Würzburg: Königshausen & Neumann.

Frank, E., N. Heinzer, and B. Tschofen. 2016. *Wolfsmanagement als kultureller Prozess. Working Paper zum Symposium "WOLFSMANAGEMENT: WISSEN_SCHAF(F)T_PRAXIS"*. Zurich: ISEK – Department of Social Anthropology and Cultural Studies (University of Zurich). http://www.isek.uzh.ch/dam/jcr:a314eaf0-c108-4f24-b2be-5c8705539eed/Working%20Paper_Auftaktsymposium.pdf.

Fudge, E. 2017. "What Was It Like to Be a Cow? History and Animal Studies." In *The Oxford Handbook of Animal Studies*, edited by L. Kalof, 258–278. New York, NY: Oxford University Press.

Gisler, P. 2016. "Key Moment. Multiple Realities of an Artefact in an Ethnographic Study of Animal-human Relations in the Zoo." *Continent* 5, 93–101.

Kirksey, S. E. and S. Helmreich. 2010. "The Emergence of Multispecies Ethnography." *Cultural Anthropology* 25, 545–576.

Korte, B. and S. Paletschek. 2009. "Geschichte in populären Medien und Genres: Vom Historischen Roman zum Computerspiel." In *History Goes Pop. Zur Repräsentation von Geschichte in populären Medien und Genres*, edited by B. Korte and S. Paletschek, 9–60. Bielefeld: Transcript.

Kowaleski Wallace, E. 2006. *The British Slave Trade and Public Memory*. New York, NY: Columbia University Press.

Latour, B. 1993. *We Have Never Been Modern*. Cambridge, MA: Harvard University Press.

Löw, M. 2016. *The Sociology of Space. Materiality, Social Structures, and Action*. New York: Palgrave Macmillan US.

Macdonald, S. 2002. *Behind the Scenes at the Science Museum*. Oxford: Berg.

Macdonald, S. 2006. "Exhibitions of Power and Powers of Exhibition. An Introduction to the Politics of Display." In *The Politics of Display. Museums, Science, Culture*, edited by S. Macdonald, 1–24. London: Routledge.

Muttenthaler, R. and R. Wonisch. 2010. "Oberfläche und Subtext. Zum Projekt 'Spots on Spaces'." In *Seiteneingänge. Museumsidee & Ausstellungsweisen*, edited by G. Fliedl, R. Muttenthaler, H. Posch, and E. S. Sturm, 77–114. Wien: Turia + Kant.

Niewöhner, J., E. Sørensen, and S. Beck. 2012. "Einleitung. Science and Technology Studies – Wissenschafts- und Technikforschung aus sozial- und kulturanthropologischer Perspektive." In *Science and Technology Studies. Eine sozialanthropologische Einführung*, edited by S. Beck, J. Niewöhner, and E. Sørensen, 9–48. Bielefeld: Transcript.

Poehls, K. 2012. "Ankommen. Wie Migration in der nordöstlichen Ägäis verräumlicht wird." In *Museum^x. Zur Neuvermessung eines mehrdimensionalen Raumes*, edited by F. von Bose, K. Poehls, F. Schneider, and A. Schulz, 89–100. Berlin: Panama Verlag.

Poliquin, R. 2008. "The Matter and Meaning of Museum Taxidermy." *Museum and Society* 6, 123–134.

Rogoff, I. 1997. "Deep Space." In *Projektionen. Rassismus und Sexismus in der Visuellen Kultur*, edited by A. Friedrich, B. Haehnel, and C. Threuter, 52–60. Marburg: Jonas.

Schäfer, R. 2015. *Tourismus und Authentizität. Zur gesellschaftlichen Organisation von Ausseralltäglichkeit*. Bielefeld: Transcript.

Schneider, F. 2012. "Die Sammlung als räumliche Praxis. Das Beispiel der volkskundlichen Sammlung von Adolf Schlabitz." In *Museum^x. Zur Neuvermessung eines mehrdimensionalen Raumes*, edited by F. von Bose, K. Poehls, F. Schneider, and A. Schulz, 138–151. Berlin: Panama Verlag.

Scholze, J. 2004. *Medium Ausstellung. Lektüren musealer Gestaltung in Oxford, Leipzig, Amsterdam und Berlin*. Bielefeld: Transcript.

Siegmundt, J. 2015. "Eine 'grüne Hölle' im Zoo. Die Regenwaldhalle als idealisiertes Habitat in Europa." In *Zoo*, edited by J. Ullrich, 81–91. Berlin: Neofelis.

Tschofen, B. 2013. "Vom Gehen. Kulturwissenschaftliche Perspektiven auf eine elementare Raumpraxis." *Schweizerisches Archiv für Volkskunde* 109, 58–79.

von Bose, F., F. Poehls, F. Schneider, and A. Schulz. 2012. "Die x Dimensionen des Musealen. Potentiale einer raumanalytischen Annäherung." In *Museum^x. Zur Neuvermessung eines mehrdimensionalen Raumes*, edited by F. von Bose, K. Poehls, F. Schneider, and A. Schulz, 7–17. Berlin: Panama Verlag.

Research material

Field notes, meeting with S. Beutler in Zurich. 9 September 2016.

Field notes, meeting with S. Beutler in her workshop. 28 September 2016.

Field notes, phone call with the President of the Foundation Council of the Domleschgian Museum. 18 November 2016.

Inspection record, Museum of Nature, Olten. 8 May 2016.

Inspection record, Dählhölzli Zoo, Bern. 14 September 2016.

Inspection record, Natural History Museum of Fribourg. 28 September 2016.

Inspection record, Natural History Museum of St. Gallen. 29 November 2016.

Interview with A. and D. 7 November2016.

Interview with the Director of the Natural History Museum of Fribourg. 6 December 2016.

Interview with the Director of the Natural History Museum of St. Gallen. 20 January 2017.

Participant observation record, hiking tour with A. and D. 15 to 17 July 2016.

Participant observation record, hiking tour with CHWOLF. 8 to 9 October 2016.

7 Modes of Involvedness. Theorising different ways of relating within the Swiss wolf debate

Nikolaus Heinzer[1]

"You're dead if you hit 26." An introduction to the field

> The Swiss, for example, have a wolf plan that's typical of that fantastic country. If you are a wolf and want to live in Switzerland, you're welcome and are part of a protected species. You can even kill sheep, but no more than 25.[2] You're dead if you hit 26.[3]

Italian biologist Luigi Boitani, a much-cited wolf expert in Europe, comments, thus, on the Swiss national wolf management concept in a *Spiegel* interview (Koch 2015) entitled "Who's Afraid of the Big Bad Wolf? Fears as Predator Returns to Europe". As the title of the quoted article suggests, since wolves began to 'naturally' recover their former habitats in central Europe about 25 years ago, uncertainties and conflicts exist about the way to deal with these events and their consequences. This is also the case for Switzerland, where single roaming wolves have been recorded since the mid-1990s and the first pack was officially identified in 2012, after *canis lupus* had been meticulously exterminated by the second half of the nineteenth century. By the end of 2019, at least three wolf packs were confirmed in Switzerland, with more packs expected to form in the near future.

Wolves embody wilderness. This is one of the few points on which the diverse perspectives comprised in the social field evolving around the return of the wolves seem to (more or less) concur. However, the discordance about the meaning and content of the word 'wilderness' is considerable. It is nothing new that the term is almost always normatively loaded (even when neutral objectivity is claimed or aimed at) and connected with different semantic, symbolic spheres with different content. Wilderness is longed for and feared, romanticised and normalised, sensationalised and stigmatised, idealised and managed. The Wild can be a recreational space with compensational functions; an autonomous ecosystem with its own, 'natural' logic; the realm of untamed and spiritual beings that live by non-human, 'naturally' just rules; a chaotic, uncontrollable, possibly deliberately 'implanted' force representing the doom of civilisation; or just another sector that has to be dealt with administratively.

One constant within these divergent perspectives is the assumption that the return of the wild wolves to Switzerland has a strong temporal dimension. Wolves have been eradicated – for good reasons or in an irresponsible frenzy, according to different points of view – and now come back in space and time, bringing along with them and somehow demanding a state or idea of nature that had been thought long gone. The homecoming of a formerly endemic species (the wolf as the lost son), the re-storing of biodiversity (the dream of the perfectly balanced primordial nature), the re-turn to a state of fearful uncertainty and lack of control (the coming true of the nightmare of dark and cruel times before civilisation); good or bad, wolves are old creatures from the past that have come back to life – I would argue to a different one – in the present of the twenty-first century. When Alpine rural actors picture themselves as oppressed indigenous people living in an artificially created wild and natural environment managed by urban people, these metaphors of the Alps as a sort of Reservation can be understood in this light.

The Reservation metaphor points to the conflictual character of wolf management.[4] Lines between different strata of society (mainly between rural and urban, lowland and highland, white and blue collar, powerful and politically weak) are being discursively redrawn and, in many cases, seemingly hardened. This includes not only the construction (or reproduction) of an antagonism between social classes and geographical regions but also the building up of competition between wild animals (nature) and humans and their domesticated animals and landscapes (culture) concerning their respective rights to a suitable habitat. This culminates in the fears expressed by some Alpine dwellers that their cultural heritage might be endangered and themselves expelled from their homelands by the politically promoted large wild predators. The programmatic welcoming back of animal 'homecomers' as enacted by official policies is, thereby, criticised as an emotional urge of urban populations to compensate for the (post)modern (post)industrial way of life, with all its ecological 'sins', by promoting the return of wolfish wilderness – at the cost of mountain dwellers' opinions, needs, spaces and lifestyles (Personal communication to the author by a Valais sheep breeder during Participant Observation on 20 June 2016). In line with this, discussions about wolves are led very emotionally by most actor groups; the attempts to de-emotionalise and objectify the discussions being a response to and, thus, part of the great emotionality.

Modes of Involvedness

This chapter has two objectives. Firstly, I want to look at the social conflicts within the debates that emerge around the return of wolves in Switzerland without reproducing the pervasive polarisation that divides the field into two antagonistic camps (e.g. wolf opponents vs. wolf supporters, Alpine countryside vs. lowland urban centres). To achieve this, I propose to think about diverse ways different actors relate actively to wolves and connected issues and entities in different situations. The second objective is then to theorise these ways of relating to and getting involved with the surroundings within the wolf debates by developing what I want to call 'Modes of Involvedness'. While outlining two extreme poles of Modes of Involvedness based on two interviews from my own fieldwork, I stress the dynamic, situational and, above all, strategic character of these modes of relating. In the last part of the article, I underline this point by giving further empirical examples that show how different Modes of Involvedness get entangled situationally.

One of the central arguments around which wolf debates rotate is the idea of 'affectedness'. Who is really afflicted by wolves' activities? Who is mainly and 'directly' concerned by the consequences and changes caused by the return of the wild predators? Who is not, or only indirectly? Who has practical knowledge and experience and who does not? Emotionality plays a vital role in underlining these claims. The status of affectedness often seems to be a very important issue, since the position of someone who is directly affected by wolf problematics allows for a widely acknowledged authority when it comes to legitimising their own perspective. Thus, while emotionality is caused by the (often) existential fears and hopes involved, it must also be seen as a strategic means in this context.

This being said, it is important to emphasise that arguments within wolf management rest on contesting bodies of knowledge, causing a great qualitative variety within the arguments. Emotively illustrated closeness to wolf issues (is based on and) generates differently characterised kinds of knowledge than do official state guidelines, scientific brochures or DNA analyses. As an example, try to compare the numeric, abstract and universal yet somewhat cryptic quality and tone of the results of genetic analyses from the laboratory that confirm an individual of the Italian wolf population to be the author of a sheep kill with the personal account of the loss of one or more breeding ewes that includes tales of year-long love and care, and gruesome details about the animal's condition when found by the owner. We get very distinct information that produces quite different pictures, meanings and consequent calls to action regarding wolves. They also suggest different ways of relating to wolves and Alpine surroundings.

An emphasis on the practices of relating, then, leads us to ask about different ways people relate to the wolf as an animal and as a broader topic, and to the (more or less extended) surroundings they and this predator willingly or unwillingly share. Based on a praxeological approach (as in Law and Lien 2012), I suggest that the practices of relating produce and stabilise different kinds of relations, and with them, different actors and *actants* (e.g. wolves, sheep, fences, landscapes, humans) (Latour 1993). This also means that different worlds are being enacted while relating to wolves in various manners. I will only address some of the aspects of this philosophical matter. In the following, I want to refocus on the empirical field of my ethnographic research by trying to generate an analytical approach that might help one to think about these divergent worlds and the different ways to relate to them without re-enacting or reinforcing the dividing and limiting momentum of the many dichotomies present in the field, such as the 'pros and cons' of the wolf or the differentiation between affected and not affected, or actually even between different social milieus or classes. Representing and working with these (sometimes diametrically antagonistic) categories does not lead one far when trying to understand properly what is going on in the Swiss wolf context.

For lack of a better term, I translate the German word *Involviertheit* literally to 'Involvedness' and, in the following, I will speak of different Modes of Involvedness.[5] On the one hand, this concept draws on the empirical category of affectedness sketched out above; on the other, it is influenced by the theoretical notion of (In)Vulnerabilities proposed by the German feminist Dominik Ohrem (2016). Ohrem understands a corporal interdependence between entities which are connected through a shared vulnerability as the basis and framework for the relational co-constructing of the world. The bodies of humans, animals and other organic beings are, thereby, horizontally open to the world.

> In a dynamic, processual-relational sense, openness to the world is to be understood here not as a status of passive receptivity for negative or positive impulses 'from outside', but rather in the mode of an actively enabling *responsivity* – as invitation, solicitation, provocation, animation and affordance.
>
> (Ohrem 2016, 71; translated by the author)

In line with an ontological or relational approach[6] and expanding on Judith Butler's notion of the body as a "porous boundary, given to others" (Butler 2004, 25, cited in Ohrem 2016, 81), Ohrem's (In)Vulnerabilities accentuate the role and importance of the body as a main vector and site of interaction, implying, thus, a relationality that is very much a corporal one. Involvedness may work on a less corporal and more cognitive level, as can be seen when looking at specific animals as iconic emotional mediators between humans and a global nature. What British literary scholar Graham Huggan says about the polar bear – "Its primary function today is that of an icon of vulnerability in a threatened world" (Huggan 2016, 14) – also holds true for wolves: they enable humans to relate to a natural world at large and Involve them in symbolic, yet emotionally deeply anchored ways. While the victimised polar bears are "co-opted as an ecological or anthropomorphic cipher for general imperilment – of 'the planet' or 'humanity' itself" (Huggan 2016, 14), the recovering wolf populations represent and relate people to a possibly recuperating global nature.

In this essay, I understand different Modes of Involvedness as representing different ways of actively relating to the world within the field of wolf management: relating to the animal wolf and the 'natural' environment that it enters and affects, and to the changes and problems that arise. Modes of Involvedness are strategically fabricated by actors, not given status, but always objects of active negotiation and performance. They comprise different modes of relationality, of being-in-the-world; through them, different wolves/

Figure 7.1 Taxidermied wolf and other wildlife in the Zoological Museum of the University of Zurich. Ecological concepts of Nature (such as the *ecosystem*) ascribe certain roles to different species and create reticular, globally binding relations between animals – and humans.
Picture: Elisa Frank (June 2016)

wildernesses/natures/worlds might come into being. Whereas in the empirical field, affectedness is always thought of as exclusive, I understand Modes of Involvedness as a strategic instrument, open to all actors and parties. Although certain strategies might be used more commonly within certain groups of actors, individual Modes of Involvedness cannot be unambiguously ascribed to certain social milieus.

While stressing that, in practice, different modes always coexist and mix, I want to draw the lines of two poles or ideal types of Involvedness: 1) Corporal-Radial Involvedness and 2) Global-Reticular Involvedness. I want to clarify two points to obviate some basic misunderstandings: a) by implicitly referring to the dichotomy of corporal vs. cognitive, I shall not insinuate a division of body and mind regarding the perception and experience of the world. Phenomenological standpoints (Merleau-Ponty 1987/1945), New Materialism (Ingold 2007)[7] and, not least, Donna Haraway, who makes a strong case for attending to the mud, fungi and bacteria that characterise our being and *becoming with* in the world (2008), make us well aware that the perception (and production) of our surroundings does indeed occur in the interrelation of both body and mind, humans and non-humans. The dichotomy present in the Modes of Involvedness tries to consider the narrative and representational practices of actors in the field, who usually put more weight on either corporal practice or cognitive knowledge. b) Again, the two ideal Modes of Involvedness do *not* represent and are *not* congruent with the extreme interest groups in the field. It would be inaccurate and insufficient to identify all wolf critics as rural, Alpine sheep holders and all wolf supporters as educated and urban; it would be equally so to equate wolf critics and supporters with the two proposed ideal Modes of Involvedness. All actors are Involved and Involve themselves along lines between these two extreme poles.

Corporal-Radial and Global-Reticular Involvedness: two extremes

In the following, I try to sketch out these two extreme poles based on two interviews I conducted. Accordingly, all the quotes in this article stem from the two interviews from 20 June 2016 (first example) and 20 October 2015 (second example), respectively.[8]

Corporal-Radial Involvedness

In the early summer of 2016, I went for a short field trip to the canton of Valais, to a region where, by that time, a number of sheep and goats had been killed by wolves. Two or three wolves had been around the area for a couple of years, and the prospect of two of them mating and forming the by then third Swiss wolf pack was imminent and confirmed later in the year. In the previous year, actions had been taken by the local sheep owners to connect with each other and with the cantonal authorities, and attempts were made to introduce herd guarding measures. These attempts failed, more sheep were killed and, as a result, the local summer pastures were permanently abandoned, which meant a big emotional loss for the sheep owners and local residents.[9] In a short 20-minute interview, Rolf Kalbermatten, one of the Valais blacknose sheep breeders implicated, spoke about some of what happened from his point of view, before taking me around to respective spots in the area for the better part of the day – after all, we were not there only to talk!

"Something I criticise", he told me in the interview, "is that at the FOEN [Federal Office of the Environment] judgements are made from the office desk. They just say, 'Build fences'. That's easy, sitting at a desk. But it's a lot of work. And you have to walk and carry all the fencing stakes." This, he explained, was a topographically problematic task when "the descending gradient is at over sixty percent". Kalbermatten then went on to describe the actual fencing work:

> I calculated it roughly. You stick the stakes in the ground, then you run through with three litz wires. So, to build the fence, you run through four

times, and four times back. All in all, that summed up to eighty kilometres.[10] [...] And in the end, you still have ten percent of loss in the herd
and the value of the meat is bad as well.

Enlarging upon the herd guarding measures, he continued:

The problem is that the ones responsible for herd guarding, respectively,
the FOEN, have been there for twenty years to see how herd guarding
works. And every year there's something new. Every year you have to
take new measures. This tells me that even they are powerless.

When asked his opinion about the then current case of the legally authorised
removal of a wolf in the area, he admitted that this might increase the
motivation to go on for a bit.

But I mean, it's not the solution. If you shoot one, there's still two left. To
do it right, I'd say you'd have to shoot all three of them. Then you have a
wolf-free area. But, yeah, if you've really got an animal that's going for
the sheep every night and you shoot it, that does give you some air to
catch your breath.

We then talked about what the situation would be like in 10 years.

Well in ten years ... That depends strongly on the politics. If there
were to be a trend reversal, we might be able to get a grip on it [the wolf
situation] again. If it keeps going on like it has in the past two years,
in ten years, a whole lot of hamlets and pretty places will be covered
with bushes.

Kalbermatten then ventured that if people in the big agglomerations were
confronted with the damage caused by wolves as he was, they might soon
change their minds. "You see it with the case of the mute swan: it's just there
[in urban areas] and defecates on some streets and people go: 'We can't accept
this!' And we here must tolerate [the wolf]." This led us to speaking about
what my interview partner stated as the biggest problem: the feeling of being
politically patronised. "Look at the president of this Swiss Pro Wolf Group
[sic],[11] who comes here and tells us, 'You've got to do it like this and like
that'. He doesn't have any idea about any of this! Not a clue. But he just says,
'it should be done this or that way'."

I asked the sheep breeder if he had ever met this person. "No, I've had e-mail
contact with him, but I don't know him personally. But that's where I have certain reservations: What does he want to tell us about if he doesn't know, if he's
not on site?" Shortly after, we concluded the interview to see for ourselves.
"Alright, let's go up to the place and have a look at it."

Figure 7.2 Sheep corral in the Valais (for spring and fall pastures). Building and maintaining the fences in the steep landscape requires physical effort and affects daily routines.
Picture: Nikolaus Heinzer (June 2016)

Corporal-Radial Involvedness is deduced from a corporally immediate experience of the local surroundings on an everyday basis. The experience of the world, in this case, of things, beings and events related to sheep, fences, state authorities and wolves, is connected in a radial way to the body: experiences, knowledge and the resulting truths and worlds come into being within and are circumscribed by the edge of the specific radius of an immediately corporal perception of the surroundings. Note the strong presence of vocabulary referring to (haptic) body functions in the statements quoted: One needs to "get a grip" on things. And one is thankful for anything giving "air to catch your breath". These being common expressions, they seem to gain a specifically organic, physical connotation in this context. The body which stands at the centre of this radial world experience is mostly the actor's own human body. But the body of the livestock can take that place, for example, when seen as potentially threatened or devaluated. The consequent actions that have to be taken in response to this (herd guarding measures) are experienced and represented in the form of physical work, bodily effort and time cost. They obtain a meaning within spatial and temporal proximity and by unsettling familiar patterns and routines. In other words, the creating of

Corporal-Radial Involvedness can be observed when claims are made of being affected by wolves and having to deal with problems *here and now*. First-hand experience and local knowledge are the relevant factors, while distanced and theoretical approaches are assumed invalid and, therefore, discredited. It is this specific, immediately corporal form of relating to the environment that is often being promoted as a close and authentic one and the only 'true' connection to nature. It is, in many cases, coined with the label of direct affectedness and asserted as the most valid access to wolf issues. However, other ways of relating to these issues exist.

Global-Reticular Involvedness

In the Autumn of 2015, I conducted an interview with the just mentioned President of the Gruppe Wolf Schweiz (Wolf Group Switzerland), David Gerke. We met in a café in Solothurn, a small town at the foot of the Swiss Jura Mountains. This interview lasted over 90 minutes and we did not have the opportunity to go out 'in the field' to discuss the topic in a more pragmatic context, as was the case with the sheep owner. Similar to the previous example, the quotes I picked out from this interview amount to what are merely small scraps of our whole conversation. I will try to contextualise them roughly. Furthermore, while they are often separated by longer phases, I will keep them in chronological order.

We had already been talking for 40 minutes when our conversation was interrupted by a call on Gerke's phone from a colleague from a nature conservation organisation. After they had exchanged information concerning the latest wolf news, we resumed our dialogue. Answering a question regarding where wolves could find their place in Switzerland, the self-proclaimed "political voice of the Swiss great predators" (Gruppe Wolf Schweiz 2017) explained the engagement of his Wolf Group with the following thoughts about what wilderness is and is not:

> At a time with seven billion people on the planet and over eight million in Switzerland, and a Europe that is a total *Kulturlandschaft* [e.g. cultural landscape], it is not the goal of species conservation to preserve nature only in areas of wilderness but to preserve nature in an integral way, including the spaces where humans are. [...] And, for me, the wolf, like the fox and other animals, is a synanthropic species. The wolf doesn't need wilderness! [...] For one, just because I look beyond the border, where I see that wolves survive in cultural landscapes in all of Europe. Switzerland is not the first country to have wolves in a cultural landscape. It's rather the last country. [...] And so, I've got to say, following the experience and the observation that the wolf is capable of surviving in these landscapes, being the only landscapes left in Europe, and seeing that there are means and instruments for people to protect livestock, to secure coexistence with the wolf, I plead in favour of the wolf being allowed to survive in cultural landscapes.

After having talked more about his general and personal understanding of nature as wilderness or cultural landscape, my interview partner repeated that 'real' wilderness no longer existed in Europe, at least not on a large scale. Notwithstanding this, he stated that "globally seen, there are probably areas in which large-scale ecosystem processes still function, without man getting involved". I interrupted briefly to make sure I had understood his meaning of natural processes as being untouched by humans in the right way. His answer was somehow ambivalent.

> Yes and no. Of course, in my opinion, man is part of nature as well. [...] Man has, like the wolf, a right to hunt, because he too is part of nature. But a natural process is, in my view, one in which man doesn't disturb unnecessarily. That doesn't mean that man is not allowed to appear in it, but that he leaves nature to its course as far as possible.

Further on, we came to discuss the images and notions people have of 'the wolf'. Gerke referred to different ways that people identify with these animals in an attempt to try to explain the general fascination wolves have for humans.

> The wolf is a pack animal, a family animal. That's certainly intriguing, and there are certain similarities in the social structure and behaviour between wolves and humans. Then there's the long history that wolf and man share. I mean, dogs are probably the oldest domesticated animals of all. [...] Man has been working with the wolf for longer than with any other domesticated or wild animal. This means the whole relationship with dogs also enters the scene here. [...] And then, of course, the wolf symbolises – for opponents and supporters – freedom, but also a lack of control, wilderness. An animal that lives freely and uncontrolled in this country, doing what it wants, that can't even be shot because, in a sense, it is so free, means something extremely positive to certain people: 'Finally, a free animal! I want to be free like that!'

Towards the end of our interview, we focused on the role science and scientific ecological knowledge play in the context of the arguments with wolf opponents. Wolf opponents' continuing efforts to define the Swiss wolves as wolf–dog hybrids were readily dismissed by the President of the Wolf Group Switzerland by referring to international genetic expertise rebutting these claims.

> And also the IUCN [International Union for Conservation of Nature], an international conservation organisation, of which Switzerland and even the national hunting association are members and which is the only international professional institution, doesn't consider the Italian wolf population or others as a hybrid population. No one does this!

Before going on about the importance of the "scientific documentation and validation of arguments", I interjected a question about the rise of the wolf opponents' political voice, which my interlocutor answered in the following way:

> The problem has broadened. At first, when we had wolves coming solely in summer, killing some sheep, maybe getting shot every now and then, only sheep owners were affected. By having packs and areas with permanent wolf presence all year long now, we have the situation that wolves come down in the winter and might come near settlements, thereby triggering new fears.

We then resumed the topic of the scientific underlining of debates, to which, according to Gerke, officials and state institutions were quite open. "[Scientific legitimisation] plays a role in influencing the building of public opinions. It is known from North American and other acceptance studies that scientific research and statements concerning wolves lead to more acceptance."

The understanding of a globally connected network-like, reticular natural world, as described and referred to in the interview, is central to the establishment of a Global-Reticular Involvedness. Relatedness to and, thus, concernment by the wolves' return is claimed through the notion that, despite the spatial distance of a few hundred kilometres that might lie between wolf territories and oneself, nature works by relating all beings on a general, ecosystemic level. Man is included as part of this global nature. Wolves embody a nature that is only locally recovering lost ground, but is really globally hit by climate change and, thus, affects all to the same degree. The spatial and temporal horizon of this Mode of Involvedness is on a far higher scale, in the sense that the distance between cities and mountains and the time span of a human life is nothing compared to the epochal dimensions in which nature evolves, making the resulting changes in human perception and use of the Alps a sometimes difficult and painful, but, all in all, acceptable consequence of 'natural' processes. Moreover, a closeness to wolves is constructed on this chronological macro level by stressing the longstanding and strong network that exists between and unites man, dog and wolf. Even though the connections between man, wolf and nature are not tangible as such, they are perceived as relevant, binding and real.

Entangling Modes of Involvedness

As mentioned above, these two ideal Modes of Involvedness represent the poles of a continuum of specific and situated cases. In the two previous examples, I have tried to pick out samples that line up relatively clearly with the two ends of this continuum. I do not say these two people or even their general views could be reduced to one or the other mode. Had I picked other parts of the two interviews, a very different – maybe even contrary – picture would have been shown. This is clearly yet more evidence that the two Modes of Involvedness are enacted very situationally and almost always combined with each other. Let us then have a look at a few examples of how these two poles are being entangled with each other in different situations and practices.

In-between positions: professional Involvedness

Rangers and other wolf management practitioners who deal with wolf issues on an immediate and everyday basis are obliged to do so through their professional responsibilities, which, in turn, are based on national and international jurisdiction. Moreover, these actors are rooted in institutions that tend to work with universalistic approaches to nature and its management. The case could be made here for a sort of professional Involvedness, which combines elements of both modes in a specific way, and which often drags professionally Involved individuals into (tangible) conflicts, bringing them into positions 'in-between front lines'. Having to implement laws, public endeavours and political assignments grounded on Globally-Reticularly Involved notions and positions and carrying them through in situations where Corporal-Radial Involvedness is often predominant, professionally Involved actors such as gamekeepers are not only very strongly involved, but also Involved in a very complex way.

Figure 7.3 Valais gamekeeper checking on a camera trap
Picture: Nikolaus Heinzer (April 2016)

In spring 2016, a wolf roamed the Lötschental valley for several weeks, killing a respectable amount of game and, thereby, catching the attention of both residents and regional and national media. I visited Richard Bellwald, the gamekeeper of the valley, and accompanied him for two days towards the end of this period. He talked about his experiences during that time[12] in an interview in 2017 for a museum exhibition broaching the issue of the return of wolves to Switzerland.

> It is a new situation and we have to learn to cope with it. In those three months, I counted more than twenty red deer, seven roe deer and also chamois that had been killed by the wolf. The hunting community and the farming community – the farmers themselves – were naturally not too pleased with my attitude and told me that I should shoot it. I had the impression that sometimes there were even vigilante groups going around at night in the hope of meeting the wolf and shooting it. However, through my presence at night, I could signal, 'No, we will not solve this problem – if it even is a problem – by using guns.' If it were an animal that was sick or injured in some way and behaved differently as a result, we would certainly react and capture the animal. But as long as that is not the case, we keep to the guidelines issued by the government and treat the animal as the law foresees.

The 'new situation' caused by the appearance of a wolf is one in which the gamekeeper is somewhat torn between two sets of interests, two sets of expectations and obligations: to protect the game, on one hand, and the wolf, on the other. Both objectives are at least jeopardised by varying but altogether difficult circumstances. How, then, does he relate to this setting? Disclosing that he is unable to protect the bodies of 'his' deer turning into corpses and reporting that he had to guard the wolf from being shot at night by means of his own physical 'presence', the gamekeeper demonstrates a strongly Corporal-Radial Mode of Involvedness with this situation. At the same time, his actions' ultimate 'guideline' is the national jurisdiction, which, in turn, builds heavily on internationally recognised reticular notions about wildlife and its management and ecology at large. In order to legitimise his doubly questioned standpoint, Bellwald has to find a way to argue in both directions, thereby referring to both Corporal-Radial and Global-Reticular Involvedness. Each one of the modes seems to be adopted for a certain (rhetoric) effect aimed at the respective critic. The objective is to add legitimacy to his own position. One can note the strong reference to the power of Corporal-Radial Involvedness to acquire the status of authenticity and closeness. At the same time, in order to reinforce one's own authority, use is made of the universal strength of Globally-Reticularly characterised wildlife management law.

Another case of professional Involvedness can be observed when paying attention to the ecologist who coincidentally joined in the conversation with the President of the Swiss Wolf Group with the telephone call mentioned above. About a week after the interview with Gerke, I met her at her organisation's regional headquarters in Basel in October 2015 for a proper interview. My first question was how she came across the wolf in her daily (work) life:

> Yeah, well, I come across it all the time, because I am the person responsible at the organisation for great predator policy and hunting policy. The wolf revealed itself to be kind of the key subject of that area. That kind of happened, on the one hand, because I am personally interested in the topic and I wanted to do this. On the other hand, there is the fact that it is simply a big medial subject matter which we as a conservation organisation are asked about and have to comment on. So, this means, [I encounter the wolf] daily, I spend seventy, eighty percent of my work time with the wolf.

Again, we notice the recurrence of Corporal-Radial modes of relating to the wolf when the ecologist states that she spends most of her work time with it, and even when she notes that her interest in the subject is, at least partially, of a 'personal' nature. At the same time, the presupposed *a priori* responsibility of her organisation to address the wolf topic suggests that the notion of a globally committing ecological network (e.g. ecosystem), connecting humans and nature on a broad level and obliging nature conservation to act, underlies the ecologist's argument.

Weakening fronts and contextualising poles

It is a sensitive issue to try to pluralise a binary opposition. The attempt to do so, even the mere identification and nomination of it, seems to automatically reinforce any dichotomy. Earlier, I asserted that the two Modes of Involvedness are not congruent with the two predominantly perceived positions of wolf opponents and supporters. Having said that, it should be repeated that the diametric opposition in question is not only a strong and somewhat inevitable discursive element in the field under research but also has an empirical foundation and validity. As poles, these two extreme positions generally attract and accumulate more of one Mode of Involvedness than the other. Even when they do not personally own sheep, wolf opponents stress their direct affectedness (e.g. by emphasising that they live where the wolves are) as a main argument to give their voice more weight and, thereby, resort strongly to Corporal-Radial ways of relating to the topic. It is ecologists and biologists who most frequently highlight the ecological role of wolves and the need and responsibility of a modern society to support their recovery. To say this does not mean that these tendencies are fixed constellations. They are not. But to totally ignore the extreme poles would be to throw the baby out with the bathwater.

To take my argument still further, I would like to touch briefly on two more ethnographic samples where the previously prevailing correlations between certain political positions and respective Modes of Involvedness are being more radically reassembled and even inverted. Firstly, take the case of a public film screening event in the Grisons Arena, a local agricultural centre in the mountainous canton of the Grisons. The film was a French home production made by a French sheep farmer documenting "The drastic consequences of the wolves' return to France" (Lecomte 2016). After the screening, inspired by the deterrent pictures from France and the daunting reports of invited German sheep holders, the Swiss wolf opponents shared their critique of current wolf policies in a dialogue with representatives of the Grison Farmers' Union, the German and French guests, and the general public at the event. The aim of the evening was to formulate a strong and clear political message that control of the Swiss wolf population should be of the highest priority to prevent the projected scenarios of the downfall not only of small livestock agriculture but of rural traditions and Alpine values.

This fear is also reflected in the front credits of the German translation of the film that was made by the event organiser (an organic farmer from the Grisons) for the Swiss screening to explain the origination of the German-dubbed version:

> In the canton of the Grisons/Switzerland, we are directly affected by the wolf's return. In summer, great Alpine pasture areas are being used by cattle, sheep and horses: an ancestral indigenous culture for centuries, even millennia. We are observing the intensifying problems with wolves in Italy, France and other European countries with great concern.
>
> Protected by the Convention of Bern, the wolf spreads rapidly unhindered [sic]. It represents an acute danger not only to our Alpine pastoral economy but also to free-range animal husbandry across all of Europe.

In addition to the immediate identification with the problems arising through the use of the first person in the plural and the 'direct affectedness' that is evoked outright in the first sentence, two aspects come forward that take this statement beyond the typical Corporal-Radial mode of relating to the wolf: firstly, a temporal extension and, secondly, a spatial expansion of their own concern. The present-day agricultural practices are being rooted in "centuries, even millennia" of cultural tradition and, thus, raised to a much higher evolutionary level. This "indigenous culture" works at the same speed as natural processes. Moreover, the endangered socio-economic complex is put into an international context and a network of relations with similar practices and cases to give the problem more relevance. In this way, the agriculturists are not only affected in a "direct" and radial manner but also on a much larger geographical scale.

In the last few years, a general trend to appropriate ecological terms, discourses and frameworks for the argumentation against the return of wolves can be traced. It also appears in the poster issued for the event. In a slogan-like fashion, the poster phrases the organiser's standpoint as follows: "For biodiversity with a natural grassland-based agriculture and pasture farming – Against the unchecked reproduction of the wolf in agriculture" (VWL 2017). The wolf is seen as a negative factor for an endemic environment that is not only pictured as a traditionally used cultural landscape but also as a flourishing eco-system created and maintained in balance by Alpine agricultural and pastoral efforts. In addition to its role as a negative socio-political impact for the region, the wolf that impedes prevailing pastoral practices, thus, turns into a more abstract ecological nuisance for the Alpine ecosystem as such.

The increasing adoption of this kind of Global-Reticular Involvedness by wolf opponents seems to indicate the strategic power it holds in the context of the wolf debates. On the other hand, many actors, such as nature conservation and wolf organisations, which, in the empirical field, are not associated with a direct affectedness and confronted with the reproach of having no legitimate link to the wolf issue, attempt to counteract these accusations by establishing physical clo-seness to the predator. For this purpose, a number of expertly guided excursions and field trips to wolf territories and dog-guarded sheep herds are offered on a regular basis by NGOs and other groups. We shall have a closer look at one offer by a politically rather wolf-neutral, small freelance hiking tour provider, called "*Wildout Naturerlebnisse*" [Wildout Nature Experiences].

Figure 7.4 Finding wolf faeces on a guided hiking tour
Picture: Nikolaus Heinzer (July 2016)

The two founders of the company, a hiking guide and a nature educationist, developed a hiking tour called "Hiking tour on the tracks of the wolves in Switzerland" (Wildout Naturerlebnisse 2017) shortly after the first Swiss wolf pack had formed in 2012 in the Calanda area at the border of the cantons of the Grisons and St Gallen. The three-day tour takes the participants straight through "the wolf's living room" (quote recorded during Participant Observation on the hiking tour on 15 July 2016), with a sleepover in an inn located a few hundred metres 'as the crow flies' from the location of the supposed wolf den, and a second night in a cabin of the Swiss Alpine Club. While no active wolf tracking is done, wolf faeces can be found and the omnipresent if improbable eventuality of a wolf sighting adds a suspenseful touch. Multiple stops are included in the trip at which various pieces of information regarding wolves and their behaviour are transmitted and shared. On one of these occasions, for instance, a wolf profile, uniting relevant wolf data, is handed to the participants in the form of a handy-sized, shrink-wrapped fact sheet that can be taken along whenever one returns to a wolf territory. The climax and implicit highlight of the tour is the encounter with a local sheep owner and his dog-guarded flock of sheep. The hikers can have a first-hand experience with livestock guardian dogs and can ask the farmer and his wife questions about how they deal with the wolves in their daily life. This exchange takes place on the pasture in the midst of the sheep and the barking livestock guardian dogs and cumulates with the farming couple inviting the tour participants to a snack at their farm nearby where they can go into the face-to-face conversation in more depth. A distinctively immediate and authentic experience is created, providing for a potential Corporal-Radial Involvedness with the Swiss wolves.

Conclusion: dynamising wilderness and engaging with practices of relating

The return of the Wild to Switzerland, as embodied by wolves, brings with it discussions, debates and conflicts that outgrow the immediate context of small livestock problematics. Wolves' migratory activities dynamise human concepts of space, nature and 'wilderness', forcing people to reconsider socio-spatial systems and relationships. While the wolves' presence is being held accountable for substantial and unbearable social and economic changes in rural Alpine areas and, consequently, for the rural exodus by some, and welcomed as an incarnation of a recovering ecosystem by others, front lines between interest groups are being (re-)drawn and existing discourses of antagonistic urban–Alpine dynamics reactivated. When wolves begin to kill sheep close to human dwellings, when fierce livestock guardian dogs trained to protect sheep from wolf attacks intimidate tourists, when issues of (livestock and human) safety become more pressing, clear (spatial and conceptual) distinctions between nature and culture seem to be what is at stake, on both a worldly everyday life basis and an abstract theoretical level.

The concept of different Modes of Involvedness tries to allow for a perspective that does not focus on or stop at these existing trenches and front lines but rather engages with strategic emotional, discursive and performative practices of positioning oneself in a certain way in relation to wolves, their actions and their effects on the surroundings, these practices always creating different wolves, different natures and different worlds. Conflicts within the Swiss wolf management context arise not merely between different interests. Divergent stances are based on different sets of knowledge, different natures and wildernesses, and different Modes of Involvedness. Looking at the distinct natures that are being enacted and the different logics of relating they call into play appears helpful when trying to understand the conflictual and emotional character of Swiss wolf debates. While I am yet unsure to which degree actors in the field are really conscious of these divergences, I believe that different Modes of Involvedness within today's Swiss Alps are interesting from a scientific, heuristic point of view: They help us to understand the complexity of the research object better. However, they might also shed light on communicative blind spots and possible roots of misunderstandings, thus, ideally providing starting points for more actively intervening and practical approaches to the topic.

Notes

1 The author writes his dissertation in the context of the research project "Wolves: Knowledge and Practice. Ethnographies on the Return of Wolves in Switzerland" (project leader: Bernhard Tschofen; project staff: Nikolaus Heinzer and Elisa Frank; project number: 162469) funded by the Swiss National Science Foundation (SNSF): http://www.isek.uzh.ch/de/popul%C3%A4rekulturen/forschung/projekte/drittmittelprojekte/wolf.html [accessed 23 November 2017]. The author wishes to thank his field partners for their time and sincerity, Philip Saunders for the proof-reading, as well as Michaela Fenske, Bernhard Tschofen and Elisa Frank for their contentual feedbacks and conceptual inputs not only to this paper.
2 As of 2017, the quantity of sheep kills allowed has been reduced to 15. Only kills in enclosures that are officially legitimised as correctly protected according to state guidelines or labelled unprotectable are considered for the count.
3 If not indicated otherwise, all the quotes in this chapter (those of my interviewees as well as cited articles and literature) that are not in English in the original have been translated by the author.
4 While wolf management, as a specific derivation of the US American paradigm of Wildlife Management, is used as the official term for the institutionalised monitoring and administration of wolves, in my research, I understand wolf management in a broader sense as including individual, everyday life and non-institutionalised ways of dealing with this animal.
5 Alluding to Latour's Modes of Existence (Latour 2013) was not my object. While Latour's earlier work leading to his Actor-Network Theory (Latour 1993) is a reference for my general theoretical approach, my claims here, obviously, are located on a much lower ranged theoretical level and point in a slightly different direction than Modes of Existence. In this chapter, I write Involvedness as well as the verbal forms with capitals to point out the use of the theoretical notion as opposed to the common verb 'to involve'.

6 I understand the *ontological turn* as an umbrella term uniting various and sometimes quite differing anti-essentialist, generally co-constructivist and relational,
process- and emergence-oriented approaches to understanding the complexities of
social (and natural) worlds and their entanglements. Latour's Actor-Network
Theory (especially Latour 1993, but also Latour 2006 and Mathar 2012) being my
personal genetic entry point to this rhizome, the philosophy of Deleuze and
Guattari (1987), the *multispecies ethnography* swarm (Haraway 2008; Kirksey and
Helmreich 2010; Tsing 2012), Mol's notion of the *Multiple* (2003), Viveiros de
Castro's *Perspectivism* (2005) and Blaser's *Political Ontology* (2009) constitute
central elements of my individual ontological/relational theoretical assemblage.
7 Without reducing him to New Materialism, Tim Ingold importantly addresses the
role of materials and materiality in the world's perception.
8 As mentioned in note 3, they have been translated into English by the author.
9 In summer 2017, the sheep were brought back onto the summer pastures, this time
under the protection of a first-hand wolf-experienced Romanian shepherd and his
livestock guardian dogs. A successful experiment, since wolves were indeed kept
from killing sheep.
10 As my interview partner himself mentioned, the procedure described refers to a
type of fence that is not officially defined as a herd guarding fence. In official herd
guarding, mostly lighter and handier electrified flexinet fences are used.
11 My interview partner here refers to the Gruppe Wolf Schweiz (Wolf Group
Switzerland).
12 "Der Wolf ist da. Eine Menschenausstellung" ("The Wolf Is Here. An Exhibition
about People") was developed by the Museum of the Alps in Bern in close coop-
eration with the SNSF project "Wolves: Knowledge and Practice" and ran from 13
May to 1 October 2017: https://www.alpinesmuseum.ch/en/exhibitions/biwak/biwa
k-19. The interview referred to here was conducted by Michael T. Ganz, transcribed
by Elena Lynch and translated into English by Pauline Cumbers. It was published in
German in the brochure accompanying the exhibition (Alpines Museum der
Schweiz and Universität Zürich – ISEK 2017, 23).

Bibliography

Alpines Museum der Schweiz (Hächler, B.), and Universität Zürich – ISEK (Tscho-
fen, B.), eds. 2017. *Der Wolf ist da: Eine Menschenausstellung*. Bern: Alpines
Museum der Schweiz.
Blaser, M. 2009. "The Threat of the Yrmo: The Political Ontology of a Sustainable
Hunting Program." *American Anthropologist* 111, 10–20.
Butler, J. 2004. *Undoing Gender*. New York, NY: Routledge.
Deleuze, G. and F. Guattari 1987. *A Thousand Plateaus. Capitalism and Schizo-
phrenia*. Minneapolis, MN and London: University of Minnesota Press.
Gruppe Wolf Schweiz, 2017. "Die politische Stimme der Raubtiere in der Schweiz."
http://www.gruppe-wolf.ch/index.php [accessed 23 November 2017].
Haraway, D. 2008. *When Species Meet*. Minneapolis, MN: University of Minnesota
Press.
Huggan, G. 2016. "Never-ending Stories, Ending Narratives: Polar Bears, Climate
Change Populism, and the Recent History of British Nature Documentary Film."
In *Affect, Space and Animals*, edited by J. Nyman and N. Schuurman, 13–24.
London and New York, NY: Routledge.
Ingold, T. 2007. "Materials against Materiality." *Archaeological Dialogues* 14, 1–16.

Kirksey, S. E. and S. Helmreich 2010. "The Emergence of Multispecies Ethnography." *Cultural Anthropology* 25, 545–576.

Koch, J. 2015. "Who's Afraid of the Big Bad Wolf? Fears as Predators Return to Europe." *Spiegel.* 30 April. Luigi Boitani interview. http://www.spiegel.de/internationa l/germany/luigi-boitani-on-the-return-of-the-wolf-to-europe-a-1028351.html [accessed 23 November 2017].

Latour, B. 1993. *We Have Never Been Modern.* Cambridge, MA: Harvard University Press.

Latour, B. 2006. "Über den Rückruf der ANT." In *ANThology*, edited by A. Belliger and D. J. Krieger, 561–573. Bielefeld: Transcript.

Latour, B. 2013. *An Inquiry into Modes of Existence: An Anthropology of the Moderns.* Cambridge, MA: Harvard University Press.

Law, J. and M. E. Lien. 2012. "Slippery: Field Notes in Empirical Ontology." *Social Studies of Science* 43, 1–16.

Lecomte, B. 2016. "Les lourdes conséquences du retour du loup." https://www.you tube.com/watch?v=3u1khQWeq2I [accessed 23 November 2017]; German-dubbed version. Available at https://www.youtube.com/watch?v=sT_2iv3QwtE [accessed 23 November 2017].

Mathar, T. 2012. "Akteur-Netzwerk Theorie." In *Science and Technology Studies*, edited by S. Beck, J. Niewöhner, and E. Sørensen, 173–190. Bielefeld: Transcript.

Merleau-Ponty, M. 1987/1945. *Phenomenology of Perception.* London and New York, NY: Routledge.

Mol, A. 2003. *The Body Multiple: Ontology in Medical Practice.* Durham, NC: Duke University Press.

Ohrem, D. 2016. "(In)Vulnerabilities: Postanthropozentrische Perspektiven auf Verwund-barkeit, Handlungsmacht und die Ontologie des Körpers." In *Das Handeln der Tiere. Tierliche Agency im Fokus der Human-Animal Studies*, edited by S. Wirth, A. Laue, M. Kurth, K. Dornenzweig, L. Bossert, and K. Balgar, 67–91. Bielefeld: Transcript.

Tsing, A. 2012. "Unruly Edges: Mushrooms as Companion Species." *Environmental Humanities* 1, 141–154.

Viveiros de Castro, E. 2005. "Perspectivism and Multinaturalism in Indigenous America." In *The Land Within: Indigenous Territory and Perception of the Envir-onment*, edited by A. Surallés and P. García Hierro, 36–74. Copenhagen: IWGIA.

VWL (Verein zum Schutz der Weidetierhaltung und ländlichem Lebensraum der Kantone Glarus, St. Gallen und beider Appenzell), 2017. "Soll es dem Wolf im Kanton Bern an den Kragen gehen?" https://www.vwl-ost.ch/termine/ [accessed 23 November 2017].

Wildout Naturerlebnisse, 2017. "Wanderung auf den Spuren der Wölfe in der Schweiz." http://www.wildout.ch/wandern-und-trekken/woelfe-am-calanda.html [accessed 23 November 2017].

Research material

Interview with Valais Sheep Breeder. 20 June 2016.

Interview with the President of Gruppe Wolf Schweiz (Wolf Group Switzerland). 20 October 2015.

Interview with a Member of a Swiss Nature Protection NGO. 26 October 2015.

Participant Observation Record, Valais Sheep Breeder. 20 June 2016.

Participant Observation Record, Wolf Hiking Tour. 15 to 17 July 2016.

8 Diverging worlds of biodiversity and biosecurity

The presence of wolves in a Swiss Alpine territory

Ilona Imoberdorf and Rony Emmenegger

Introduction

> The wolves that are walking back into a landscape are doing so less as predators on still-present livestock but, rather, as heralds of a newly reinvigorated naturality, one in which the competing rhetorics of biosecurity and biodiversity curiously intertwine.
>
> (Buller 2008, 1587)

The return of wolves since the mid-1990s has polarised the debate on the possibility of human and wolf coexistence in Alpine Switzerland. Whereas for some, wolves are considered dangerous beasts, for others, they are persecuted creatures worthy of protection (Hunziker and Landolt 2001). Scholarly literature on the subject explains the negative or positive attitudes to wolves differently, as the result of a diffuse anxiety (Linnell et al. 2002; Røskaft et al. 2003) or fascination and their vernacular perpetuation (Dingwall 2001; König 2010); of the economic or emotional anguish involving conflicts of use (Boitani 1995; Wallner and Hunziker 2001); of the spatial and social distance to territories with a presence of wolves (Karlsson and Sjöström 2007; Wallner and Hunziker 2001); or of divergent value-determined attitudes among socio-demographic classes (Wild-Eck and Zimmermann 2001) and among socio-economic milieus (Caluori and Hunziker 2001). Hunziker, Hoffmann, and Wild-Eck (2001, 321) pointed out that the acceptance of predators can only be understood in relation to "the attitudes towards the wildness and preferences with respect to landscape development" (authors' own translation; see also Wallner and Hunziker 2001). The spatial imagination of regions with a presence of wolves exerts, thus, considerable influence on the various positions adopted towards wolves and is decisive for determining whether wolves are perceived as being *threatened* or *threatening* within the regions in question.

In this chapter,[1] we examine diverging conceptions of Alpine nature and culture regarding their political consequences for humans and wolves (see also Swyngedouw 2007, 19–20; Uggla 2010, 79). To position oneself 'for' or 'against' the presence of wolves requires the description of the Alpine nature influenced by the presence of wolves. Assuming a political ontology lens,[2] this involves less the representation of a single reality 'out there' but far more the ways in which an Alpine nature is evoked by way of descriptions thereof (Blaser 2013; Descola 2013; Descola and Palsson 1996). Various adopted positions produce, in turn, differing worlds than conceived as specific Alpine geographies. Hence, the debate on the presence of wolves in Switzerland does not turn primarily on the collision of "more or less accurate cultural representations" (Blaser 2013, 551–552) of a singular Alpine reality. As with various other environmental conflicts (Blaser 2009; de la Cadena 2010; Li 2013, 400), the debate is rather, to cite Fabiana Li (2013, 400), the expression of "struggles over the enactment, stabilization and protection of multiple socionatural worlds".

Our analysis focuses on the debate on the presence of wolves in Alpine Switzerland, exemplified by the Alpine semi-canton Oberwallis.[3] Based on a qualitative analysis of the positions taken by various interest associations in this debate,[4] we analyse the embeddedness of wolves in Alpine geographies as a product of socio-political struggles. We evince that the acceptance of wolves is essentially dependent on whether and by whom a region with a presence of wolves is conceived or enacted as 'wild' or 'civilized'. We further demonstrate that Oberwallis is both a material *and* an imaginary object of diverging worlds in which "rhetorics of biodiversity and biosecurity", as Henry Buller (2008, 1587) puts it, "curiously intertwine" (see also Lowe 2015; Ojalammi and Blomley 2015, 53). The contradictory appearance of wolves as a threatened or threatening species in Oberwallis, thus, appears as a result of the tension between these "competing philosophies of nature" (Buller 2008, 1583) that constitute diverging Alpine worlds in which humans and animals are put into place – or put into question. We argue that the implementation of Oberwallis as a territory, as a "bounded meaningful space" (Delaney 2005, 15), is decisive in the debate between advocates and adversaries of wolves in determining *who* may legitimately speak *for whom* within the region's territorial boundaries.

Our description of diverse – in part corresponding, in part colliding – conceptions of culture–nature and human–animal geographies provides a substantial contribution to a nuanced understanding of the polarised debate on the presence of wolves in Switzerland. The present chapter provides a five-step examination of this debate. The first part consists of a historical outline of the changing relationships and encounters between human beings and wolves. In the second part, we discuss how the return of the wolves to Oberwallis is framed as an urban–rural conflict. In the third part, we describe various assumptions about the hunting behaviour of wolves and the ramifications of this for biodiversity and, respectively, biosecurity in Oberwallis. Part four focuses on possible measures taken towards the presence of wolves which, in different worlds, appear legitimate for *governing* wolves. In the fifth part, we discuss the constitution of Oberwallis as a political entity whose population is resolutely against the presence of wolves and who should be politically represented, respectively. We conclude with a summary of the political consequences the debate on wolves in Oberwallis has for human beings and animals *being* in the world.

Alpine contact zones

Wolves were exterminated in various European countries as a consequence of their changing relationships with humans during the nineteenth century (Breitenmoser 1998; Etter 1992). Wolf populations came increasingly under pressure, above all, due to the changing habitat in the wake of industrialization, as this led to the deforestation in Alpine areas as sources of energy (McShane and McShane-Caluzi 1996) and to the increased agricultural exploitation of forests as forest pastureland (Küster 1995). For wolves, the concomitant degradation of forests as a natural habitat for wild animals resulted both in deprivation of domains of retreat and in a decline of their potential prey (Breitenmoser 1998, 284). Attempts by wolves to satisfy their hunger with farm livestock have, in turn, led to their being more rigorously tracked down by the existentially affected populations. Targeted by the latter, the population of wolves in Switzerland further came into the sights of increasingly efficient and organised hunting. New types of rifle, for example, improved the accuracy of range and weather resistance in shooting. Snap traps or poisoned baits were used for combating wolves in addition to the hunting drives (KORA 2005, 9). Similarly, the introduction of state-approved bounties on proof of wolves killed impacted negatively on their population (Etter 1992).

While wolves had been exterminated in Alpine regions by around 1900 (Breitenmoser-Würsten et al. 2001, 15), the populations in the various regions of Alpine Europe subsequently began to recover and increase. This development is linked to changing forms of land usage and growing efforts in environmental protection (Rüegg et al. 2002, 141). Furthermore, a trend towards species protection also turned wolves into subjects of legal coding in Switzerland during the second half of the twentieth century (see also Ojalammi and Blomley 2015): as part of the 1979 Bern Convention (Appendix II), for example, wolves were listed as a strictly protected species of animal and, by way of Switzerland's ratification of this Convention, were awarded protected status in 1982. This legal status of 'the wolf' has remained valid until today.[5] From a historical perspective, the legal protection of wolves corresponds to a certain paradigm shift in the "biopolitical imperative": from biosecurity to biodiversity (Ojalammi and Blomley 2015, 53). To date, 'the wolf' continues to be protected in two additional legal texts which stipulate that hunting the animal is prohibited: the Swiss Hunting Law[6] and the Hunting Regulation.[7]

Wolves began returning to southern Alpine regions in the early 1990s. Coming up from Northern Italy (Boitani and Ciucci 1993; Ciucci et al. 1997), single animals were first sighted in France in 1992 (Buller 2008, 1588) and in Switzerland in 1995 (Breitenmoser 1998, 284) – especially in the Oberwallis region. Since then, the wolf population has been growing steadily in Switzerland, with Oberwallis, as the port of entry to Switzerland for wolves from Italy, especially exposed. Musiani and Paquet (2004, 56) see the possibility for the coexistence of human beings and wolves as generally lying in "wild and semiwild areas". Accordingly, Glenz et al. (2001, 60) have modelled the potential habitats of wolves in Oberwallis, concluding that such areas partly persist at mid-range altitudes: whereas anthropogenic activity in lower locations is too high, which makes coexistence correspondingly impossible, higher altitude habitats are prohibitive owing to a lack of prey and inhospitable, geomorphological conditions. Yet, different complementary studies emphasise wolves' advanced abilities for adaptation, especially in areas with a relatively high population density (Rüegg et al. 2002, 143).

The question of potential coexistence is debated from biological or ecological perspectives, whereas numerous killed goats and sheep testify to overlapping territorial claims of human beings and wolves in Alpine regions. In the summer of 2011, wolves in Oberwallis killed a total of 130 sheep, as Urs Zimmermann, the gamekeeper of canton Wallis, explained in a telephone interview with the first author of this chapter. High numbers of kills may be explained, above all, as a consequence of the long absence of large predators, making traditional protective systems for small livestock redundant (Linnell et al. 2002; Røskaft et al. 2003; Treves and Karanth 2003). With the absence of wolves, a specific form of summering has established itself in Oberwallis, which has made it possible to maintain small-scale, individualised sideline enterprises of so-called 'open pasturage' in the Alps. In case of wolf attacks, the owners of such livestock often take their livestock into the valley as a precautionary measure, discontinuing the summering of the livestock partially or fully. In several cases, by way of an emergency measure, a shepherd with sheepdogs is dispatched to the Alpine meadow in question. The significance of killed livestock is not necessarily of an economic nature – especially in cases in which the system of state compensation functions for breeders – but may well be far more of an emotional kind (Wallner and Hunziker 2001, 195–200; see also Kruuk 2002; Sommers et al. 2010). The presence of wolves creates a threat situation, especially for livestock farming and the breeding of the typical blackface sheep, which is highly esteemed in Oberwallis and contributes considerably to the regional identity. The local importance of blackface sheep is exemplified by the fact that the debate on the wolf presence in Oberwallis circulates primarily around the threat to sheep, while the threat to the typical blackneck goats is rarely addressed.

The overlapping territorial claims of human beings and wolves in Oberwallis pivot on the origins of human–animal geographies that are discussed in this chapter. Our analysis does not focus primarily on the question whether wolves enter 'natural' or 'cultural' terrain by returning to Oberwallis or whether the ecology in Oberwallis permits coexistence. We understand nature and culture rather as spatial constructs, the significance and delimitation of which result from the encounters of human beings and animals (Hobson 2007, 251; Kirksey and Helmreich 2010, 546). It is by virtue of the ongoing – seldom physical, but predominantly imaginary – encounter between human beings and wolves in Oberwallis that human–nature and human–animal geographies are meaningfully constituted. According to this perspective: "wolves are geographers too, enacting space through forms of mobility and territoriality" (Ojalammi and Blomley 2015, 56). In a *more-than-human world*, wolves appear as geographers, not, for example, because they actively position themselves in the debate about their presence in Oberwallis but because they provoke a socio-political dispute between people simply through their presence. Between advocates and adversaries of wolves reacting upon the presence of wolves, it is then the dispute about the appropriate treatment and about the implementation of the requisite measures – shooting or protection – which constitutes the subject of the debate to be outlined in the following.

Territorial threats

> [Wallis] is especially exposed to the problems that have emerged with the return of the wolves since 1995. In spite of its experiences, it has until now sought in vain to sound the alarm. Today, it has been proven that wolf packs live on the borders of Wallis.
>
> (KWJV 2011)

The return of wolves to Alpine Switzerland has become the object of severe accusations in the debate between the adversaries and advocates of wolves. For the former, various inconsistencies are grounds for scepticism. According to one breeder association representative, it is at least striking that while wolves had been absent in Switzerland for several decades, they have now suddenly reappeared in various locations. No less remarkable is the fact that wolves are referred to as pack animals in general discussions, whereas in Oberwallis, atypically, one only finds single animals. The representative of the Association for the Defence against Large Carnivores (VVG) further cites various stories claiming that wolves are anaesthetized, driven to the Swiss border and abandoned whilst still sedated, before being "lured into Switzerland" by means of "traces of sheep blood". Such stories would be plausible if wolves, atypically in view of their actual behaviour, entered Switzerland for the most part via pass roads. A representative of the conservationist association World Wide Fund for Nature (WWF 2019), however, sees no contradiction in this pattern of mobility, emphasising that pass routes, for both humans and wolves, simply facilitate the least difficult mountain crossings. An anonymous representative also stresses, somewhat irritated, that wolves have not arrived "on four wheels", and that even the most vehement wolf adversaries ought to have recognised this. Indeed, the overlapping of wolves' mobility patterns with human infrastructures in various contexts has been proven (Ojalammi and Blomley 2015, 56). However, the question about the origins of the return of the wolves along these paths remains open to interpretation and is a matter of controversial discussion in Oberwallis.

   In the debate on the presence of wolves, conservationist associations are generally under suspicion of having actively promoted the return of wolves to Oberwallis (Caluori and Hunziker 2001, 176; Wallner and Hunziker 2001, 196–197; see also Stokland 2013, 2). According to one representative of the

Wallisian Association for Wildlife Biology, such a suspicion among wolf adversaries is based on early experiences connected with the return of lynxes to Switzerland. Following the first sighting of lynxes 40 years ago, the return of these animals was initially also presented as a 'natural' process, beginning with the search for 'natural' ways of entry. This then became problematic, because those conservationist associations and governmental departments actively participating in the repopulation concealed their involvement, which they admitted only after a 10-year statute of limitations had passed. Today, this presentation is as good as uncontroversial and documented in expert circles (Breitenmoser 1983; Breitenmoser and Breitenmoser-Würsten 1990). It also, as the director of the Centre for Herd Protection Jeizinen (HSZJ) explains, substantiates a certain mistrust "in peoples' minds" in the controversy surrounding the return of the wolves to Oberwallis. Articulated by wolf opponents, this presentation hardens the suspicion of similar 'opaque' involvement by various institutes in the return of wolves and weakens the credibility of conservationist efforts in Oberwallis significantly.

The involvement of conservationist associations and their efforts to protect the wolves in Oberwallis is seen by many as an expression of an urban–rural conflict. According to the director of the HSZJ, the conflict about the presence of wolves is substantiated essentially by the fact that the conservationist associations WWF and Pro Natura, each of which have their central offices in larger cities, such as Zurich and Basle, mobilise their members primarily in the wolf-friendly urban milieu. As he explains, in the eyes of particularly city dwellers, wolves are seen as survival artists epitomising a lost freedom that, in view of accelerated social developments and the "rapidity of work", has fallen by the wayside. Opposed to these are the "[local farmers] all of whom have their niches, and who, alongside their professions, attend to the animals almost every weekend or in the evenings" – for these, this is "pure nature, and now someone [the wolf] is imposing on their territory". In his view, unfortunately, conservationist associations have showed themselves insufficiently sensitised to events in Oberwallis during the wolves' return and the outset of the discussion; consequently, each of their activities achieved little on the ground. For him, this then explains why these associations were perceived by many in Oberwallis as a "Swiss affair" and were put on the "hit list of wolf opponents" in the region.

Clear reference is shown in scholarly literature on the subject to the fact that rurally influenced regions, which are less accepting of wolves, contrast with urban areas, with a greater acceptance of wolves. Wallner and Hunziker (2001, 195) show in their analysis of the social acceptance of wolves in Switzerland that social and spatial distance to areas with a presence of wolves influences the attitudes of indifference or enthusiasm significantly. Behr, Ozgul, and Cozzi (2017, 1925) in their socio-demographic analysis, account for the fact that "people living at higher elevations", where the presence of wolves is more likely, tend to position themselves *against*, and "people living in densely populated areas" more *for* the presence of wolves. In spite of such a 'statistical tendency', there are two reasons why we find it difficult to take this acceptance map for granted: firstly, the geographic specification in given, clearly prescribed outlines specifically ignores the *spatially situated* production of Alpine geographies and, thus, the overlapping of multiple, diverging and conflict-charged worlds which collide in the debate on the presence of wolves. Secondly, the articulation of a given acceptance map contributes significantly to the polarisation of the debate as it reifies a rural–urban dichotomy in a bordered, political landscape.

Thus, the debate on the presence of wolves puts into effect the mobilisation of an urban–rural dichotomy and allows for conservationist interests to appear in Oberwallis as an undermining of local autonomy (see also Wallner and Hunziker 2001, 191). After all, the Oberwallisians would doubtless not introduce crocodiles into Lake Thun (a lake close to the capital city Bern), as the representative of the Oberwallisian Goat Breeding Association pointed out in a discussion with the first author of the present chapter: "[Imagine] yourself going for a swim when suddenly a crocodile came your way. 'This is absolutely mad', you'd say." As this representative figuratively intimates, the presence of wolves in rural-influenced, mountainous country makes as much sense as a crocodile in a lake in Switzerland's urban-influenced lowland country. The representative of the hunting association Diana Brig (DB) further criticised "a certain tendency" among city dwellers "to classify Oberwallis as an open park for wolves and bears" and "to see our rural region as their holiday resort". According to the representative of the VVG, the Oberwallisians have always tracked the wolves, whereas city dwellers are simply fascinated by them. In this connection, critical voices argue that advocates have a predominantly glorified romantic notion of wilderness and find a bogeyman in the notion of the "postmodern wolf proponent" (Caluori and Hunziker 2001, 179) that, with this wilderness or wild nature, seeks to maintain the pristine, fascinating, sublime and beautiful counter-world to the civilised life in an urban and industrial environment.

From the point of view of the wolf opponent, the presence of wolves is incompatible with the culture economy in Oberwallis. In the view of Stremlow and Sidler (2002, 28–29), cultivation and utilisation of a region exert substantial influence on aesthetic perception, which explains the negative attitude towards wolves. In this sense, Wallner and Hunziker (2001, 197) have shown that "acceptance towards predators in nature" becomes less, "the stronger somebody characterises a region shaped by human beings as nature" (authors' own translation; see also Bauer 2005, 111, 130–131). Consequently, the presence of wolves contradicts the definition of the role of human beings as masters over nature, who assert their superiority in the constant struggle against wilderness (see also Stremlow and Sidler 2002, 28). In other words, for the "traditional wolf opponent" (Caluori and Hunziker 2001, 176), the presence of wolves represents the return from the wilderness symbolically and is, thus, incompatible with the hegemony of the human being over "the tamed natural world". Wolves constitute, as the VVG (2010) states in a flyer, a fundamental threat to grazing sheep in a nature conceived as a cultural region and the cultivation of "our meadows and Alps".

The possible consequences of the return of wolves are described by way of various scenarios (see also Bauer 2005, 111), underscoring the situation of threat it poses to Oberwallis: many fear that owing to the presence of wolves, summering must be spatially and structurally limited or even discontinued altogether. Similarly, the discontinuance of small livestock breeding is assumed to result in a reduced demand for the products and services of agricultural suppliers. Some, furthermore, assume that the mistake of open pasturage for sheep threatened by wolves leads to regrowth. According to the VVG representative, such a spread of "dense vegetation, impenetrable even for wild animals" would, in turn, lead to an increased danger of avalanche, since snow on the long, non-grazed grass would slide more easily than it would on grazed surfaces. He further highlights extensive fires to be reckoned with as a consequence of natural regrowth, which would be disastrous for the affected area; since "the helicopter is unable to reach this altitude to extinguish the fire with water". With the return of wolves in Oberwallis, the wilderness threatens to spread dangerously and to destabilise the region fundamentally as it affects both the cultivated Alps and the economic relations maintaining the Alpine landscape. It is anticipated that neglected Alps will then no longer conform to the aesthetic ideas of tourists and that the latter will, thus, cancel their visits to Oberwallis. In Oberwallis, as a region in which the tourist industry represents a substantial source of income for the local population, such a scenario is assumed to have severely negative consequences.[8]

According to the opponents of wolves, the latter threaten not only the Alpine culture economy but also no less the human beings in the valleys. Breeder associations, arguing that wolves come closer to the densely populated areas, above all in winter, enhance the anxiety about wolves, for instance. It also holds for Oberwallis that "the threat is always there as a potential" (Buller 2008, 1591) – all the more so since there have been no reports of wolves attacking human beings. The increased potential of danger and the anxiety about possible wolf attacks on human beings becomes plausible in that single animals had been sighted in Oberwallis close to settlements in the autumn of 2011. However, the sighting came as little surprise to Urs Zimmermann, gamekeeper in Oberwallis, particularly because he understands wolves as animals "capable of adaptation", and who follow their prey, game and livestock down into the valley in autumn. In a telephone interview with the first author of this chapter, he even referred to a call from a worried mother who had heard about the sighting and felt it necessary to make inquiries of the gamekeeper as to whether it was still safe to take her children with her when attending to her cattle.

Enabling biodiversity, endangering biosecurity

Competing philosophies of nature intertwine curiously in the debate on the presence of wolves in Oberwallis. From the wolf-critical perspective, the presence of wolves poses a threat to the life of domestic animals and to spaces dominated by human beings. A "pervasive wolf terror" exists about the senseless and bloodthirsty wolf attacks, which threaten to "massacre grazing sheep on the Alps" (*Walliser Bote* 2011b). As the representative of the Blackface Sheep Breeding Association explains, wolves take one Alp after the other in Oberwallis that is "densely populated by sheep and wild game" – presenting a comprehensive threat to the biosecurity of the entire region. Here, wolves not only seem invariably unpredictable, but also appear to be tactically operating actors. Such tactical behaviour has also been confirmed by the director of the HSZJ when explaining that wolves alternate their hunting grounds and wait until their prey in a given area have calmed down and become less attentive, following which they then later return to the same valley basin.

From a wolf-friendly perspective, the presence of wolves represents less of a threat to the existing culture economy, but far more an opportunity for the existence of Oberwallis as a functioning ecosystem. In a brochure, for instance, the WWFS (2011) emphasises the role of the wolf as a "health policeman", ensuring a fit and healthy wildlife capable of regulating the population of wild animals. Their legitimation at the top of the food chain (see also Buller 2008, 1589) is based on their ability to select, fundamentally questioning the human being in his function as a hunter – as will be discussed later. According to the WWF (2019) brochure mentioned, the return of wolves signifies the "return of a natural enemy" for wild animals. The consequence of the presence of wolves is that wild animals behave in accordance with their species, become cautious and spread themselves better. This effect of the presence of wolves on the behaviour of wild animals consequently reduces browsing damage to young trees, whereby forests are better able to rejuvenate and "formerly suppressed species" once again make their appearance. The presence of wolves is further assumed to support other predators, which benefit from the remains of wolves' kills. For the WWF, there is no doubt about the ecological value and utilisation of the presence of wolves: the return of wolves is a "godsend" as it guarantees the healthy population of wild animals and biodiversity.

Wolves and human beings compete for hegemony over Alpine nature in the debate on the presence of wolves in Oberwallis. According to conservationist associations, hunters are not capable of competing with wolves for two reasons. Hunters only hunt during the hunting season, which makes all-year round selection impossible. As a hunter, states the Wallisian Association for Wildlife Biology (fauna.vs 2005) in an information piece, the human being does not concentrate primarily on the old and the infirm, as it hardly makes sense as a challenge in the context of trophy hunting (see also Dahles 1993). Representatives of hunting associations, however, do not agree with such a stylisation of wolves. They see wolves more as large predators that hunt according to criteria entirely different from those used in hunting by humans and incomparable with those of the moral standards of the latter group. The representative of the Oberwallisian Hunting Association is convinced: the wolf "knows no moderation but the hunter does". Accordingly, DB (2012) legitimates the hunt in the protocol of its annual assembly by pointing to the social task with which it has been appointed, namely, to nurture and regulate wild game: "In contrast to the large predators, such as the lynx and wolf, for many decades, the hunter, as regulator, understands how to approach natural resources according to the principle of sustainability". The Cantonal Wallisian Hunting Association (KWJV 2011) also places emphasis on the careful planning of the hunt, which enables hunters, as "guarantor[s] for biodiversity", to meet their obligations "lock, stock and barrel". In the same statement, the association further praises the creation of hunting reserves by local hunting associations (*Dianas*)[9] for facilitating the protection of wild animal populations, from which numerous Wallis animal species would then stand to profit.

For the most part, opponents of wolves share the view of hunting associations and, similarly, emphasise the superiority of 'the hunter' over 'the wolf'. One of the fundamental criticisms of wolves is formulated by way of reference to a civilised hunt as this has developed in the absence of wolves. With their extinction, one can learn on the VVG (2011) homepage, hunting in all civilised countries has reached a higher and more peaceful evolutionary stage concomitant with a waning of animal cruelty. This may be seen, for example, by the fact that animals hunted by hunters and not by wolves are mostly killed immediately. In contrast to the presentation of hunters as trophy collectors, the term 'selection' is described as an inherent aspect of the 'civilised' hunt, in which the shooting of older and sick animals is the real moral duty. As part of the 'civilised' hunt, hunting reserves are also seen as a modern achievement: hunters determine where wild animals are to be protected from the hunt and where nature is allowed to be, indeed, should and must be 'wild'. In a hunting reserve conceived along these lines, hunters assume hegemony over nature and protect nature through the legal exclusion of human influences within its territorial boundaries. For these reasons, the representative of the VVG is convinced that hunting in 'civilised countries' should remain in the hands of authorised hunters. In his view, hunting in Oberwallis appears to be an integral part of the civilising process, over the course of which, the human being asserts his hegemony over nature.

In Oberwallis, as many opponents of wolves are convinced, wolves are to be found in an environment inappropriate to the species and, consequently, have no reason to exist. This apparent fact is expressed and confirmed in the thesis of the 'abnormal' wolves, whose behaviour is entirely different from those conspecifics in the wild (see also Buller 2008, 1586). One learns on the VVG (2011) homepage that attacks on sheep are unethical and torturous, especially because prey mortally wounded by wolves may take hours to die: wolves attack the stomach region after having first incapacitated the animal by biting. Wolves may not only rip out the intestines of a living expectant sheep but occasionally even the embryo. The wolf bites: "excessively, senselessly, and ecologically without use" (VVG 2007). In other descriptions similarly outlined, the scene of the kill – the Alpine meadow – has the characteristics of a dangerous wilderness and is, thus, enacted as a dangerous location. In the act of killing, metaphorically speaking, the wild entity of the wolf colours the scene of the event and poses a threat to the biosecurity on Alps.

Nobody contests the substantial damage wolves have caused in past years by killing grazing sheep in the debate on the presence of wolves in Oberwallis. No less uncontested is that wolves have repeatedly killed and attacked more sheep on different Alps then they could eat on-site. The assumption that wolves are possibly worked up into a kind of bloodlust based on a 'natural' drive to hunt seems plausible for representatives on both sides of the debate. One representative of the Wallisian Association for Wildlife Biology explains such a scenario in the following manner: the wolf kills a farm animal very 'normally'. But then its hunting instinct is triggered in such a way that it also begins attacking other animals in the immediate vicinity; however, instead of killing these, it simply incapacitates them. Similarly, another representative of a conservationist association draws on the biology of wild animals when seeking to explain that the high number of kills results from a stockpiling principle. Consequently, whereas the high number of kills appears to be the result of 'natural' behaviour, the opponents of wolves tend to view this as an expression of 'abnormal' behaviour. The VVG representative, for example, holds that in the case of wolves, such as the one which once killed 90 sheep in Unterwallis (Val Ferret), this was probably a so-called *Spinnerli*, or maniac, namely, a degenerate wolf. In principle, he would concur with a biological and rational drive to stockpile, but this for him by no means suffices to explain why this could result in the killing of 90 animals.

No less uncontroversial is the fact that the presence of wolves in Oberwallis and the current practice of summering in 'open pasturage' are incompatible. According to the representative of DB, wolves can only live out their "killer drive" on a herd of sheep, which is not the case with wild animals, since these are available in fewer numbers and their behaviour is different. As a "simple shepherd" noted – in keeping with that of the VVG representative – when attacked, wild animals will flee up steep slopes too strenuous and, thus, prohibitive for the wolf. Regarding sheep, by contrast, as the director of the HSZJ explains, a herd instinct has been trained such that in the case of an attack, they tend rather to remain together than flee. The VVG representative confirms this observation in his description of the reaction of sheep under attack: "Head in, butt out". In his view, this behaviour is not 'natural', but useful for human beings. Accordingly, the numerous wolf kills may be explained as resulting from domestication and the 'unnatural' behaviour of domesticated sheep when under attack by wolves. What the VVG representative still considers 'natural', however, is the close relationship between mother ewes and their lambs, which when attacked, do not leave their offspring behind and invariably return to them rather than fleeing.

Managing wolves

Different measures for handling wolves appear legitimate in diverging worlds. For the opponents of wolves, the shooting of wolves is a necessary countermeasure against the threat situation and to guarantee biosecurity in Oberwallis. Indeed, the shooting of wolves in Switzerland, organised in accordance with the legal coding of 'the wolf', is possible but complicated.[10] The cantons are permitted to issue a shooting licence in consultation with an inter-cantonal commission when the criteria stipulated by the Federal Department for the Environment (BAFU) are fulfilled: (1) a defined number of killed livestock by a 'damage-causing wolf' (2) within a designated perimeter and (3) during a predetermined period of time can be proven (BAFU 2010).[11] Kills by wolves within a perimeter with a recognised presence of wolves are determined on this basis, however, only if (4) reasonable protective measures were taken (BAFU 2010). Consequently, producing evidence of the presence of a 'damage-causing wolf' entails quantifying damage to life and proving the presence of wolves as a question of biosecurity.[12] If a shooting licence has been issued, it is then the prerogative of the cantonal game-keeper to realise the shoot within a predetermined period of time. In addition to this temporal limitation stipulated in the shooting licence, the validity is limited to the spatial perimeter within which the damage to livestock has been registered, so as to guarantee that the specific 'damage-causing wolf' is shot.

Thus, biodiversity and biosecurity clash within the law as competing philosophies of nature in a complex way. The legal geography in which wolves are fundamentally protected is, therefore, open to clearly defined temporal and spatial exceptions in which the life of these animals is at stake. In the debate on the presence of wolves, however, the relevant laws are usually equated with the protection of wolves and, thus, necessarily questioned by opponents of wolves. As the VVG (2007) criticizes, the attitude of the BAFU towards wolves and its philosophy of harmonious coexistence between wolves and human beings is "nonsense". In a letter to the editors of the local newspaper, Walliser Bote (2011a), the objective of his criticism pivots on the conditions that must be fulfilled for the issuing of a shooting licence: The Federal Department requires "that the population of wolves is 'sufficiently large and stable'" and, consequently, requires that wolves must be first established before their regulation becomes necessary. A representative of one hunting association shares the opinion that the Federal Department has misunderstood the gravity of the situation and poses the critical question whether it would first consider what to do only after the formation of wolf packs in Switzerland.[13] In his view, a change in the treatment of wolves has long since been overdue.

By contrast, the advocates of wolves welcome the legally enshrined protection of 'the wolf' and call rather for an adaptation of the established culture economy to the presence of big predators. Herd protection especially appears as an appropriate technical means to guarantee the coexistence of wolves, small livestock and human beings in the Alpine regions. According to the director of the HSZJ, for a "proper profiling" of an Alp: "planning must be implemented to determine when and where the animals are allowed to eat [and be secured] such that these can also be shepherded". This involves not only protecting sheep from wolves but also limiting the free movement of sheep during summering. In this way, sheep in open grazing would climb up directly to the freshest grass but without eating older grass in lower areas. In his estimation, this presents a problem for two reasons: firstly, sheep would not graze on the Alpine meadows provided for them and, consequently, these areas would begin to deteriorate increasingly, despite cultivation. Secondly, sheep would compete with wild animals, for which fresh grass at higher altitude locations represents their nutritional basis. In his world, herd protection appears as an instrument used not only for enabling coexistence between wild and domesticated animals in the Alps but also as a means for consolidating existing boundaries between 'wild' and 'domesticated' spaces in an Alpine landscape.

The implementation of herd protection demands adaptation to established practices of summering, explains the director of the HSZJ. However, it is precisely this that presents the difficulty to him: "To break up one's own back garden and to contribute to a larger community is what the majority of people – to some extent for understandable reasons – are not prepared to do". Herd protection ideally comprises planning the summering and the accompanying of livestock – frequently consisting of animals belonging to several owners – by a shepherd who leaves a herd to graze at specific locations and fences off these locations overnight. Due to the small-structured, individualised sideline enterprise in Oberwallis, the implementation of this form of herd protection amounts to established summering in open grazing and is correspondingly difficult to design. One alternative in discussion for Oberwallis – as well as for other regions – is an attenuated form of herd protection whereby, as a rule, two to three herd protection dogs are integrated into the sheep herds and accompany these in 'open pasturage'. Even though this form of herd protection is only rarely implemented in Oberwallis, its advocates support it as part of responsible use of the Alps.

The discussion on herd protection as a possible measure for the establishment of a responsible use of the Alps is rife with contradictions. Many accuse sheep farmers of not carrying out summering in open grazing in accordance with specifications outlined in tradition but out of a certain convenience. As an anonymous representative explained, breeders, in summer, simply "send" sheep "off" into the Alps and summer them in open pasturage without supervision – losses accepted accordingly. In such descriptions, the actual threat to domesticated sheep comes not only from wolves, but also from their farmers, who accept falls or other causes of death among their sheep as a matter of course. In fact, only 6.2 per cent of all causes of death among sheep could be traced to large predators nationwide during the summering period of 2011 (Werder 2012, 12).[14] In view of this circumstance, various questions are raised in the debate on the presence of wolves in Oberwallis about whether the farmers ought not to assume full responsibility for the protection of their sheep themselves. According to critical voices, one must look at the present form of Alpine usages very critically indeed – above all in the context of today's substantial sheep farming subsidies – or accept the end of the Alpine culture economy and the concomitant deterioration of the Alps necessarily entailed by this.

Whereas many question small livestock breeding in Oberwallis, others see farmers' responsibility for the cultivation of the Alps already more than fulfilled: The summering of sheep and the "great efforts involved in the care and maintenance of the culture landscape and the Alpine flora connected with this" (VVG 2011) in Oberwallis are understood as a voluntary service to the general public. Here, the tradition of cultivating and managing the Alps is especially tailored to mountainous Oberwallis. Herd protection is, thus, considered an extreme and demanding adjustment to the advance of wolves in Oberwallis and is correspondingly questioned. Two negative consequences of herd protection for classic sheep breeding, for example, are emphasised. On the one hand, the sheep have less weight gain over the summer and a less superior fleece than would be the case in open pasturage. According to the director of the Blackface Sheep Breeding Association, such sheep occupied only the rear ranks in the livestock shows in the autumn. Herd protection, in his view, contradicts the symbolic position of blackface sheep in Oberwallis – since "it is actually for this that we keep them, for the autumn fair, the lamb and ram fair, exhibitions and such like". On the other hand, falls are feared when sheep flee from the herd protection dogs that live among them. Whereas wolf opponents usually present falls as a result of wolf attacks, herd protection dogs, despite their socialisation within a flock of sheep, are here equally assumed to pose a certain threat.

It seems clear for sheep protection associations that herd protection has been attempted but has failed to function "here in Oberwallis" (*Walliser Bote* 2011d). The representative of DB also complains that the relevant federal office has "repeatedly claimed that it [herd protection] does work providing it is implemented, but that the application is lagging to some extent". The VVG representative also emphasises that herd protection measures can be implemented only in part: wolves hunt intelligently by distracting the herd protection dogs or by dispatching one among the wolf pack from the other side in advance, so that the sheep run over to the opposite side where the other wolves are waiting for them. Wolves also use bad wind and weather conditions as a means to make it difficult for the dogs to protect the sheep. Here again, wolves are described as tactical actors: they play a role in the territorial struggle for hegemony in Oberwallis, in so far as they are capable of reacting to man's herd protection measures.

Threatened 'us', threatening 'others'

> So, people then say here: "What's the point; what do we want with the wolf; what are we supposed to do with it? – It lives here and eats sheep. Are we supposed to turn our lives upside down simply because one of them happens to be lurking about the place?" It's insane!
>
> (Representative of the Oberwallisian Forest Association)

The debate about the presence of wolves is an expression of a competition between the different associations about *who* may speak *for whom* within the territorial boundaries of Oberwallis. The situation is clear for many wolf opponents: legislation protects the wolves, threatens the Alpine culture economy and, consequently, constitutes a threat itself. Thus, argues the VVG representative in a letter to the editors of the local newspaper in the 1970s, the state government – above all, the Federal Council – recommended that farmers in mountainous regions shift from cattle breeding to small livestock. However, with the ratification of the Bern Convention which followed shortly thereafter, the Federal Council invited the wolf into the country "with full enthusiasm" (*Walliser Bote* 2011a), and thereby ensured that it eats the recently required small livestock again. In another letter to the editors, another opponent of wolves also shares his worry that this, as an upshot, will lead to thousands of jobs being lost, simply because "a few armchair slacktivists in Brussels and Bern seek to undermine mountain farming and tourism" (*Walliser Bote* 2011c) by all the means at their disposal. By the same token, the author of this letter also discerns a strategy in this structural change – spearheaded by economic interest associations – for the "passive restructuring of Switzerland" by means of depopulating Alpine valleys, such as Oberwallis. Following Kropp (2010, 48–51), such a malaise may be understood as resulting from the fact that 'wilderness' is more often planned for structurally weak areas, while experts and politicians in urban areas undertake decision-making. This circumstance does, in fact, frequently coincide with the disadvantages of rural areas. Regarding the debate about the presence of wolves, accordingly, it is frequently raised as part of the general threat situation for Oberwallis as it allows the presentation of the local population as 'victims' of external interests.

No less occasion for criticism is provided by herd protection measures promoted and encouraged by the state government – the Federation. The VVG representative criticises in a letter to the editors, above all, the fact that "provisional herd protection is so dramatically underfinanced that it is only just enough for intervention troops – herds and dogs unfamiliar with the respective animals – which only then intervene after the damage has been done" (*Walliser Bote* 2011a). The WWFS (2011) also recognises in a brochure that the Federation has only half-heartedly been supporting the farmers with herd protection measures, and still lacks working in close co-operation with the relevant national agricultural associations. However, lack of support in herd protection measures, according to the representative of the Wallisian Association for Wildlife Biology, is apparently not the fault of the relevant federal department but the parliament, which has failed to allocate necessary funding. Despite there being some degree of consensus between adversaries and advocates, the question about the actual responsibility for the implementation of herd protection measures remains disputed: for some, the Federation has not provided the necessary support despite the initiative of those concerned; for others, the continual lamentation over the Federation's lack of support is an expression of deficient self-responsibility among sheep breeders.

Many wolf opponents present the legal provisions for wolf protection as an example of the undermining of local autonomy. Consequently, they demand greater political support at a national level to represent the interests of Oberwallis' agriculture, especially in the capital. The VVG and the national breeder associations, for instance, demand in a joint policy paper that the cantons work to ensure that the strict legal provisions for the protection of wolves be loosened and that the Bern Convention responsible for this be annulled (VVG, SZV, and SZZV 2010). In point of fact, political advances from Wallis are highly prominent at a national level. As an example, a considerable number of the 14 motions covered in the National Assembly during the autumn session in 2010 on the treatment of wolves were introduced by Wallis politicians (Bundesversammlung 2017). The focus of these motions covered, above all, not only the loosening of laws regulating large predator shooting covered in the Bern Convention and regulatory possibilities for populations of large predators but also the strengthening of herd protection. Consequently, representatives of the breeder associations also widely share the view that the canton of Wallis, as well as the Wallis politicians, understood the threat situation and, therefore, collaborated with their associations: contrary to the Federation, the canton and cantonal politicians concerned themselves with the farmers, recognised the work of agricultural associations and shared the anxiety about farmers no longer being able to farm the canton.

The representative of DB also supports the local political commitment of cantonal politicians: "That a Wallis politician is a vehement opponent of the presence of wolves is legitimate. After all, he [sic.] represents the view of the majority. To talk politics any differently would be wrong, and neither would it make any electoral sense." According to a statement by the Cantonal Wallisian Hunting Association, Wallis can count itself lucky that "all Wallis parliamentarians voted in favour of a reasonable protection of the wolf, which accounts for the interests of agriculture, tourism, economy, environment and hunting in Wallis" (KWJV 2011). According to this association, the canton also has a mandate "to care for the health and diversity of the fauna", which is why the canton must be able to influence the population of large predators; since large predators, such as wolves, could destroy the "results of decades of efforts by the agency in support of wild animals and the hunters". Hence, from this standpoint, the guarantee of biodiversity and biosecurity is primarily a dependence on local self-determination as a necessity to regulate wolf populations in response to local demands.

Conservationist associations, on the contrary, are somewhat sceptical about the Wallis government and their politicians. Some of them argue that members of the cantonal government or cantonal politicians contribute little in the way of solutions, but endorse the shepherds' negative attitudes, for example, regarding herd protection. According to the anonymous statements of one association representative, this is taken to such extremes that even the cantonal official responsible for herd protection does not represent herd protection at the Oberwallis Chamber of Agriculture meeting. In addition, the director of the HSZJ observes a certain reluctance of the Oberwallis politicians towards the promotion of herd protection, explaining it as a result of their fear to be misperceived as advocates of wolves. In his view, however, there is little contradiction between undertaking measures regarding herd protection and an adversary position towards wolves: although one may not be in favour of wolves, one can still endeavour to do something to minimise the damages.

According to statements issued by the WWFS (2011): "large predators are well-suited for use as bogeymen" and serve as an "argument for passing the buck elsewhere for grievances". For all intents and purposes, the wolf is here said to serve primarily as a practical scapegoat for politicians collecting votes in rural regions. The Wallisian Association for Wildlife Biology (fauna.vs 2010) also expresses criticism regarding the role of politicians in a press release: Oberwallis politicians have put forward the most diverse motions and advances and liked to show themselves committed – "Wallis' federal politicians as the avant-garde". Rather than being concerned with the problems of the farmers, however, they prefer rambling on in lengthy discussions, vociferating about provisos to Article 22 of the Bern Convention and the annulment thereof. Meanwhile, more wolves have migrated and valuable time has been lost for local farmers for whom the politicians have provided no services. The annulment of the Bern Convention because of hunters' claims and grievances about open pasturage would then, according to the WWFS (2011), simply amount to making a "mountain out of a molehill". According to these conservationist associations, an international accord, such as the Bern Convention, cannot be annulled on the grounds of conflicts of use. Rather, they advocate for an understanding of the presence of wolves as a Europe-wide phenomenon; which is why unilateral action by a small region such as that of Oberwallis and the local demands for the shooting of wolves is dismissed as unreasonable.

Despite all kinds of disagreements between advocates for and adversaries of wolves, the representatives, for the most part, agree at least on two points: firstly, that the mood regarding the presence of wolves in Oberwallis is predominantly negative. Here, Oberwallis is conceived as a political entity whose population is united against the presence of wolves. The reason, as the representatives of DB points out, is that "the people who live in regions with wolves are those who experience the suffering; not the natives of Zurich or the officials of the relevant federal offices". To be sure, as the representative of the Blackface Sheep Breeder Association concedes, there are also those in Oberwallis for whom this issue is of no concern, those who are indifferent to the presence of wolves, or even those who advocate them – "there are all manner of views". However, according to a representative of the Oberwallisian Forest Association, the percentage of wolf advocates in Oberwallis is "in the thousandths", since everyone has "a hunter or shepherd in the family" and is, thus, affected by the return of the wolves. In addition, the WWF Oberwallis (2019) highlights on its homepage that it is "the natives, especially those in agriculture" who wish to prevent the return of large predators. All these representations (re-)produce Oberwallis as an Alpine territory whose population is united against ecological templates in general (Hermann and Leuthold 2003, 78) and more specifically against the presence of wolves.

Secondly, there is general agreement in Oberwallis that it would be better in the long-term not to be all too publicly in favour of the wolf or else show interest in the subject. An anonymous interview partner explained the situation in the following way:

> It is better to hold back on one's views somewhat, and not make them too public; in this way, one can then make more progress. And in certain circles, so much as mentioning the subject, as I am now, will result in their not having anything to do with you when discussing other subjects on another occasion.

Fifty or sixty years ago, as the representative of the Oberwallisian Goat Breeding Association claims, it would have been impossible to be in the region simply due to an interest in wolves – as the first author of this chapter did to conduct interviews, for instance – for this would have led to immediate expulsion from the region. In view of the enormous rage among many Oberwallisians, this representative is convinced that one may well even today anticipate conflicts: "turmoil will at some point erupt should a wolf advocate appear. He'd be hung, drawn and quartered". As this example impressively documents, a rhetoric is used in the struggle against the wilderness and its representatives that contrasts with achievements of human civilisation. However, the biting is not as bad as the barking in Oberwallis, adds the same representative: nowadays, one is more civilised and solving conflicts by recourse to violent means is not the only option available, but banging one's fist on the table is sometimes simply a must. One way or another, this concluding example, proves the ability of opponents of wolves to evoke violence as a powerful rhetorical means in the struggle over the enactment of diverging worlds. It comes along with the localization of advocates of wolves as so-called '*Üsserschwizer*' or 'non-Swiss' who hail from beyond Oberwallis' boundaries and are disavowed the rights to speak for Oberwallis and its people.

Conclusion

In this chapter, we have discussed the collision of diverging conceptions of nature and culture as exemplified in the debate about the presence of wolves in Alpine Switzerland, and shown the ways in which human beings, wolves and sheep are embedded in Alpine geographies. Following the feminist scholar, Donna Haraway (2008, 244), this debate may be understood as being a product of the encounter between wolves and humans in a "contact zone" where the very existence of humans and animals in the world is at stake. As the introductory quote indicates, wolves have broken open and destabilised the consolidated boundary between nature and culture by their return to

Oberwallis, compelling the renegotiation of human–animal geographies (see also Buller 2008, 1585). Accordingly, the presence of wolves provokes a competition between adversaries and advocates of wolves about the implementation and stabilisation of Oberwallis as an Alpine territory in which biodiversity and biosecurity intertwine curiously. In such a struggle, Oberwallis is both the material and imaginary object of diverging worlds in which human beings and animals are included or excluded and in which wolves, particularly, are either threatened or threatening. In this struggle, furthermore, Oberwallis is diversely activated as an Alpine territory in which certain measures prove legitimate for responding to the presence of wolves and for maintaining or creating biosecurity or biodiversity within. Setting territorial boundaries is thereby instrumental as it determines who within Oberwallis is awarded or denied the right to exist, and *who* may legitimately speak *for whom* within these boundaries.

Notes

1 We are indebted to Bernhard Tschofen and Michaela Fenske for their support in the process of writing our contribution to their edited volume, as well as to Stephan Hochleithner, Jevgeniy Bluwstein, Nikolaus Heinzer and Pia Hollenbach for helpful comments on earlier versions of this book chapter. We are, furthermore, beholden to Philip Saunders for translating the German manuscript of this chapter into English.

2 A 'political ontology' signifies a radical turning away from traditional western concepts of science according to which nature becomes comprehensive as objective reality by way of scientific methods. In keeping with the 'social studies of science' (see, above all, Law 2004; Law and Mol 2002; Mol 2002), the focus of political ontology lies in various descriptions of nature and the way in which it is evoked through such descriptions. For further details, see Hacking's (2002) and Smith's (2015) works on a 'historical ontology'.

3 The canton of Wallis is geographically and politically divided into Unter- und Oberwallis. The separation corresponds to a linguistic boundary: French is spoken in Unterwallis, whereas German is spoken in Oberwallis.

4 All interest associations examined in this chapter have their main or ancillary offices located in Oberwallis. The selection comprises two breeder associations (Schwarznasenzuchtverband: SN, and the Oberwalliser Ziegenzuchtverband: OZIV), two hunting associations (Diana Brig: DB, and the Oberwallis Jagdverband), three conservationist associations (the World Wide Fund for Nature: WWF, Pro Natura Oberwallis and the Gesellschaft für Wildtierbiologie: fauna.vs), as well as three other associations (der Verein zur Verteidigung gegen Grossraubtiere: VVG, the Herdenschutzzentrum Jeizinen: HSZJ, and the Oberwalliser Forstverband). Our analysis is based on data collected through interviews with representatives of the associations listed above (conducted by the first author of this chapter in the summer of 2011) and a selection of documents written or supported by them (homepages, statements, policy papers, media press releases, flyers, brochures and books). This database is complemented by a series of newspaper articles and letters to the editor published in the local press, *Walliser Bote*, as well as a telephone interview with an expert, Urs Zimmermann, the gamekeeper of canton Wallis. All this research material is written in German. Philip Saunders translated the selection of direct quotations used in this chapter to English.

5 Bern Convention; SR 0.455. *Übereinkommen über die Erhaltung der europäischen wildlebenden Pflanzen und Tiere und ihrer natürlichen Lebensräume* [Convention on the Conservation of European Wildlife and Natural Habitats], Bern, 19 September 1979.
6 Jagdgesetz; SR 922.0. Federal *Bundesgesetz über die Jagd und den Schutz der einheimischen wildlebenden Säugetiere und Vögel* [Federal Law on Hunting and Protection of Indigenous Wild Mammals and Birds], 20 June 1986.
7 JSV Jagdverordnung; SR 922.01. *Verordnung über die Jagd und den Schutz wildlebender Säugetiere und Vögel* [Regulation on the Hunting and Protection of Wild Mammals and Birds], 29 February 1988.
8 Wolves are attractive marketing items for the regional tourist industry. In 2011, for example, when the data collection was carried out, Brig-Belalp Tourismus and Visp Tourismus advertised a *Wolfspfad* (wolf path) on their home pages and decorated this with images.
9 Hunting associations are named after the patron goddess of hunters *Diana* and are geographically located (e.g. DB).
10 The conditions for shooting are stipulated in the legally binding management plan (Konzept Wolf Schweiz) of the Federal Department for the Environment (Bundesamt für Umwelt: BAFU). Since July 2015, the conditions are elucidated in the reviewed Hunting Decree, JSV.
11 According to the Konzept Wolf Schweiz (BAFU 2010), the cantons may issue shooting licences for a wolf in cases in which certain perimeters either during a period of four consecutive months in which 35 animals have been killed by a wolf or if 25 animals have been killed by the same wolf within one month. The criteria for the issuing of a shooting licence have remained substantially unchanged in the Konzept Wolf Schweiz (BAFU 2016).
12 The cantons issued a total of 17 shooting licences between 2000 and 2016. Canton Wallis issued 13 of these, while 6 of these shooting licences could not be issued in fixed periods and, consequently, expired (BAFU 2017). In addition to legal shoots, illegal shoots have also been documented (Verein CHWOLF 2017).
13 A mere six wolves were proven to be in Oberwallis in the year in which the data were collected for this chapter. In 2012, the first pack formation followed in the Calanda region (border region of the cantons Graubünden and St. Gallen). To date, two other packs have been discovered in Switzerland in the cantons of Tessin and Oberwallis (KORA 2017).
14 In addition to kills by wolves, sheep and goats die from various causes, such as lightning bolts, rockfalls, falling, sicknesses or other large predators (Werder 2012, 12).

Bibliography

BAFU: Bundesamt für Umwelt, 2010. *Konzept Wolf: Managementplan für den Wolf in der Schweiz*. Bern: Bundesamt für Umwelt.

BAFU: Bundesamt für Umwelt, 2016. *Konzept Wolf Schweiz: Vollzugshilfe des BAFU zum Wolfsmanagement in der Schweiz*. Bern: Bundesamt für Umwelt.

BAFU: Bundesmat für Umwelt, 2017. "Wolf." Homepage. https://www.bafu.admin.ch/bafu/de/home/themen/biodiversitaet/fachinformationen/massnahmen-zur-erhaltung-und-foerderung-der-biodiversitaet/erhaltung-und-foerderung-von-arten/grossraubtiere/wolf.html.

Bauer, N. 2005. *Für und wider Wildnis – Soziale Dimensionen einer aktuellen gesellschaftlichen Debatte*. Zürich: Haupt.

Behr, D. M., A. Ozgul, and G. Cozzi. 2017. "Combining Human Acceptance and Habitat Suitability in a Unified Socio-ecological Suitability Model: A Case Study of the Wolf in Switzerland." *Journal of Applied Ecology* 54 (6), 1919–1929.

Blaser, M. 2009. "The Threat of the Yrmo: The Political Ontology of a Sustainable Hunting Program." *American Anthropologist* 111 (1), 10–20.

Blaser, M. 2013. "Ontological Conflicts and the Stories of Peoples in Spite of Europe: Toward a Conversation on Political Ontology." *Current Anthropology* 54 (5), 547–568.

Boitani, L. 1995. "Ecological and Cultural Diversities in the Evolution of Wolf–Human Relationships." In *Ecology and Conservation of Wolves in a Changing World*, edited by L. Carbin, S. Fritts, and D. Seip, 3–11. Alberta: Canadian Circumpolar Institute.

Boitani, L., and P. Ciucci. 1993. "Wolves in Italy: Critical Issues for Their Conservation." In *Wolves in Europe – Status and Perspectives*, edited by C. Promberger and W. Schröder, 74–90. Ettal: Munich Wildlife Society.

Breitenmoser, U. 1983. "Zur Wiedereinbürgerung und Ausbreitung des Luchses (Lynx lynx) in der Schweiz." *Schweizerische Zeitschrift für Forstwesen* 134 (3), 207–222.

Breitenmoser, U. 1998. "Large Predators in the Alps: The Fall and Rise of Man's Competitors." *Biological Conservation* 83 (3), 279–289.

Breitenmoser, U. and C. Breitenmoser-Würsten. 1990. "Status, Conservation Needs and Reintroduction of the Lynx (Lynx Lynx) in Europe." *Nature and Environment Series* 45, 1–43.

Breitenmoser-Würsten, C., K. Robin, J.-M. Landry, S. Gloor, P. Osslon, and U. Breitenmoser. 2001. "Die Geschichte von Fuchs, Luchs, Bartgeier, Wolf und Braunbär in der Schweiz: ein kurzer Überblick." *Forest Snow and Landscape Research* 76, 9–21.

Buller, H. 2008. "Safe from the Wolf: Biosecurity, Biodiversity, and Competing Philosophies of Nature." *Environment and Planning A* 40 (7), 1583–1597.

Bundesversammlung, Das Schweizer Parlament, 2017. "Abstimmungs-Datenbank: Die Namentlichen Abstimmungen im Nationalrat." Homepage. https://www.parlament.ch/de/ratsbetrieb/abstimmungen/abstimmungs-datenbank-nr.

Caluori, U. and M. Hunziker. 2001. "Der Wolf: Bedrohung und Lichtgestalt – Deutungsmuster in der Schweizer Bevölkerung." *Forest Snow and Landscape Research* 76 (1/2), 169–190.

Ciucci, P., L. Boitani, F. Francisci, and G. Andreoli. 1997. "Home Range, Activity and Movements of a Wolf Pack in Central Italy." *Journal of Zoology* 243 (4), 803–819.

Dahles, H. 1993. "Game Killing and Killing Games: An Anthropologist Looking at Hunting in a Modern Society." *Society & Animals* 1 (2), 169–184.

DB: Diana Brig, 2012. *Protokoll Generalversammlung 2012* [Protocol of the Annual Assembly]. Protokoll, 28 January 2012.

de la Cadena, M. 2010. "Indigenous Cosmopolitics in the Andes: Conceptual Reflections beyond 'Politics'." *Cultural Anthropology* 25 (2), 334–370.

Delaney, D. 2005. *Territory: A Short Introduction*. Malden, Oxford and Carlton: Blackwell Publishing.

Descola, P. 2013. *Beyond Nature and Culture*. Chicago, IL: University of Chicago Press.

Descola, P. and G. Palsson, eds. 1996. *Nature and Society: Anthropological Perspectives*. London: Routledge.

Dingwall, S. 2001. "Ravenous Wolves and Cuddly Bears: Predators in Everyday Language." *Forest Snow and Landscape Research* 76 (1/2), 107–120.

Etter, T. 1992. *Untersuchungen zur Ausrottungsgeschichte des Wolfes (Canis lupus L.) in der Schweiz und den benachbarten Gebieten des Auslands*. Thesis/Dissertation. Zürich: Eidgenössische Technische Hochschule (ETH), Abteilung Forstwirtschaft.

fauna.vs: Walliser Gesellschaft für Wildtierbiologie, 2005. *Wolfskonzept oder Selektion Vegetarischer Wölfe* [Wolf Concept or Selection of Vegetarian Wolves]. fauna.vs Info, 30 June.

fauna.vs: Walliser Gesellschaft für Wildtierbiologie, 2010. *Wie lange wollen wir noch zuschauen?* [How Much Longer Do We Want toWatch?]. Press release, 30 August.

Glenz, C., A. Massolo, D. Kuonen, and R. Schlaepfer. 2001. "A Wolf Habitat Suitability Prediction Study in Valais (Switzerland)." *Landscape and Urban Planning* 55 (1), 55–65.

Hacking, I. 2002. *The Scope of Logic, Methodology and Philosophy of Science.* Dordrecht: Springer.

Haraway, D. 2008. *When Species Meet.* Minneapolis, MN and London: University of Minnesota Press.

Hermann, M. and H. Leuthold. 2003. *Atlas der politischen Landschaften: Ein weltanschauliches Porträt der Schweiz.* Zurich: vdf Hochschulverlag.

Hobson, K. 2007. "Political Animals? On Animals as Subjects in an Enlarged Political Geography." *Political Geography* 26 (3), 250–267.

Hunziker, M., C. Hoffmann, and S. Wild-Eck. 2001. "Die Akzeptanz von Wolf, Luchs und 'Stadtfuchs': Ergebnisse einer gesamtschweizerisch-repräsentativen Umfrage." *Forest Snow and Landscape Research* 76 (1/2), 301–326.

Hunziker, M. and R. Landolt, eds. 2001. "Humans and Predators in Europe: Research on How Society is Coping with the Return of Wild Predators." *Forest Snow and Landscape Research* 76 (1/2), 1–326.

Karlsson, J. and M. Sjöström. 2007. "Human Attitudes towards Wolves, a Matter of Distance." *Biological Conservation* 137 (4), 610–616.

Kirksey, S. and S. Helmreich. 2010. "The Emergence of Multispecies Ethnography." *Cultural Anthropology* 25 (4), 545–576.

König, B. 2010. *Die Darstellung des Wolfsbildes im Kontext geschichtlicher Entwicklungsprozesse: eine wissenschaftliche Analyse am Beispiel ausgewählter Printmedien seit 1873.* Thesis/Dissertation. Freiburg: Albert-Ludwigs-Universität, Fakultät für Forst- und Umweltwissenschaften.

KORA: Koordinationsstelle Raubtierökologie und Wildtiermanagement, 2005. *Dokumentation Wolf: erstellt im Auftrag des Bundesamts für Umwelt, Wald und Landschaft (BUWAL).* Muri: KORA.

KORA: Koordinationsstelle Raubtierökologie und Wildtiermanagement, 2017. "Situation CH." Homepage. www.kora.ch/wolf.

Kropp, C. 2010. "Wildnis morgen: Szenarien zukünftiger Wertschätzung." In Laufener Spezialbeiträge: *Wildnis zwischen Natur und Kultur: Perspektiven und Handlungsfelder für den Naturschutz.* Laufen/Salzach: Bayerische Akademie für Naturschutz und Landschaftspflege, 45–52.

Kruuk, H. 2002. *Hunter and Hunted: Relationships between Carnivores and People.* Cambridge: Cambridge University Press.

Küster, H. 1995. *Geschichte der Landschaft Mitteleuropas von der Eiszeit bis zur Gegenwart.* München: Beck.

KWJV (Kantonaler Walliser Jagdverband), 2011. *Der Wolf im Wallis* [The Wolf in Wallis]. Statement, 16 April 2011.

Law, J. 2004. *After Method: Mess in Social Science Research.* London: Routledge.

Law, J. and A. Mol. 2002. *Complexities: Social Studies of Knowledge Practices.* Durham, NC: Duke University Press.

Li, F. 2013. "Relating Divergent Worlds: Mines, Aquifers and Sacred Mountains in Peru." *Anthropologica* 55 (2), 399–411.

Linnell, J., R. Andersen, Z. Andersone, L. Balciauskas, L. C. Blanco, L. Boitani, S. Brainerd, *et al.* 2002. "The Fear of Wolves: A Review of Wolf Attacks on Human." *Norsk Institutt for Naturforskning NINA Oppdragsmelding* 731, 1–65.

Lowe, C. 2015. "From Biodiversity to Biosecurity." In *The Routledge Handbook of Political Ecology*, edited by T. Perreault, G. Bridge, and J. McCarthy, 493–503. London: Routledge.

McShane, T. and E. McShane-Caluzi. 1996. *Swiss Forest Use and Biodiversity Conservation*. WWF Case Study Edition. Gland: WWF International.

Mol, A. 2002. *The Body Multiple: Ontology in Medical Practice*. Durham, NC: Duke University Press.

Musiani, M. and P. C. Paquet. 2004. "The Practices of Wolf Persecution, Protection, and Restoration in Canada and the United States." *AIBS Bulletin* 54 (1), 50–60.

Ojalammi, S. and N. Blomley. 2015. "Dancing with Wolves: Making Legal Territory in a More-than-human World." *Geoforum* 62, 51–60.

Røskaft, E., T. Bjerke, B. Kaltenborn, and J. Linnell. 2003. "Patterns of Self-reported Fear Towards Large Carnivores Among the Norwegian Public." *Evolution and Human Behavior* 24 (3), 184–198.

Rüegg, D., P. Camin, L. Fischer, and P. Heimann. 2002. "Der Wolf im Nationalpark der Abruzzen–Erfahrungen sammeln und Vergleiche mit der Schweiz ziehen." *Schweizerische Zeitschrift für Forstwesen* 153 (4), 140–145.

Smith, J. 2015. *Nature, Human Nature, and Human Difference: Race in Early Modern Philosophy*. Princeton, NJ: Princeton University Press.

Sommers, A., C. Price, C. Urbigkit, and E. Peterson. 2010. "Quantifying Economic Impacts of Large-carnivore Depredation on Bovine Calves." *Journal of Wildlife Management* 74 (7), 425–1434.

Stokland, H. B. 2013. "Molecularising Nature: How Scandinavian Wolves Became Natural." *Forum* 16, 1–9.

Stremlow, M. and C. Sidler. 2002. *Schreibzüge durch die Wildnis: Wildnisvorstellungen in Literatur und Printmedien der Schweiz*. Zürich: Haupt.

Swyngedouw, E. 2007. "Impossible 'Sustainability' and the Postpolitical Condition." In *The Sustainable Development Paradox*, edited by R. Krueger and D. Gibbs, 13–40. New York, NY: The Guilford Press.

Treves, A. and K. U. Karanth. 2003. "Human–Carnivore Conflict and Perspectives on Carnivore Management Worldwide." *Conservation Biology* 17 (6), 1491–1499.

Uggla, Y. 2010. "What Is This Thing Called 'Natural'? The Nature–Culture Divide in Climate Change and Biodiversity Policy." *Journal of Political Ecology* 17 (1), 79–91.

Verein CHWOLF2017. "Geschichte der Einwanderung der Wölfe in die Schweiz." Homepage. https://chwolf.org/woelfe-in-der-schweiz/rueckkehr-der-woelfe/geschichte-der-einwanderung.

VVG: Verein zur Verteidigung gegen Grossraubtiere 2007. *Der "Verein zur Verteidigung gegen Grossraubtiere" stellt sich vor* [The "Association for the Defence against Large Carnivores" introduces itself]. Brochure, February.

VVG: Verein zur Verteidigung gegen Grossraubtiere 2010. *Für den Schutz der Nutztiere – gegen den Abbau der Biodiversität* [For the Protection of Farm Animals – against Biodiversity Degradation]. Flyer, 4 October.

VVG: Verein zur Verteidigung gegen Grossraubtiere 2011. "Verein zur Verteidigung gegen Grossraubtiere [Association for the Defence against Large Carnivores]." Homepage. http://lupo.tech-box.ch

VVG, SZV, & SZZV: Verein zur Verteidigung gegen Grossraubtiere, Schweizerischer Schafzuchtverband, and Schweizerischer Ziegenzuchtverband2010. *Positionspapier des SZV und des SZZV zur Problematik der Grossraubtiere* [Position Paper of the

SZV and the SZZV on the Problem of Large Carnivores]. Policy paper, 22 September.

Walliser Bote, 2011a. "Der Wolf und das Wunschdenken des Bundesamts für Umwelt [The Wolf and the Wishful Thinking of the Federal Office for the Environment]." Letter to the editor from the representative of the VVG, 26 July.

Walliser Bote, 2011b. "Terror auf den Alpen [Terror in the Alps]." Letter to the editor, 9 August.

Walliser Bote, 2011c. "Wölfe – Eine Schande! [Wolves – a disgrace!]." Letter to the editor, 10 August.

Walliser Bote, 2011d. "Schäfer halten nichts von Herdenschutz [Shepherds Don't Believe in Herd Protection]." News article, 2 September.

Wallner, A. and M. Hunziker. 2001. "Die Kontroverse um den Wolf–Experteninterviews zur gesellschaftlichen Akzeptanz des Wolfes in der Schweiz." *Forest Snow and Landscape Research* 76 (1/2), 191–212.

Werder, C. 2012. *Abgänge/Verluste von Schafen während der Sömmerung*. Birmensdorf: AlpFutur-Publikationen.

Wild-Eck, S. and W. Zimmermann. 2001. "Raubtierakzeptanz in der Schweiz: Erkenntnisse aus einer Meinungsumfrage zu Wald und Natur." *Forest Snow and Landscape Research* 76 (1/2): 285–300.

WWF Oberwallis, 2019. "Lebensräume für Pflanze, Tier und Mensch [Habitats of plants, animals and humans]." Homepage. https://www.wwfoberwallis.ch/unser e-themen/biodiversitaet/.

WWFS: World Wide Fund for Nature Schweiz, 2011. *Fragen und Antworten zum Wolf* [Questions and Answers about the Wolf]. Brochure, 23 March.

Interviews

Interview with the representative of the Association for the Defence against Large Carnivores (Vereins zur Verteidigung gegen Grossraubtiere: VVG) in the summer of 2011.

Interview with the representative of the Blackface Sheep Breeding Association (Schwarznasenzuchtverbandes: SN) in the summer of 2011.

Interview with the director of the Centre for Herd Protection Jeizinen (Herdenschutzzentrums Jeizinen: HSZJ) in the summer of 2011.

Interview with the representative of the Oberwallisian Forest Association (Oberwalliser Forstverband: OWFV) in the summer of 2011.

Interview with the representative of the Diana Brig (DB) in the summer of 2011.

Interview with the representative of the Oberwallisian Goat Breeding Association (Oberwalliser Ziegenzuchtverbandes: OZIV) in the summer of 2011.

Interview with the representative of the Oberwallis Hunting Association (Oberwallis Jagdverband: OWJV) in the summer of 2011.

Interview with the representative of the Pro Natura Oberwallis (PN) in the summer of 2011.

Interview with the representative of the Wallisian Association for Wildlife Biology (fauna.vs) in the summer of 2011.

Interview with the representative of the World Wide Fund for Nature (WWF) Schweiz in the summer of 2011.

Telephone interview with Urs Zimmermann, gamekeeper of the canton of Wallis, 16 December 2011.

9   Getting close(r)

Alive or dead: biography, individuality and agency of the wolf MT6

Irina Arnold

> Bene's is the first lot to be drawn. Number 28. He offers it to me, but I decline
> because I hope mine will also be drawn. However, it is not. Five people are
> allowed to go inside and meet the wolves. Mrs. Vogelsang, the caretaker of
> the wolves, explains: no bags allowed inside; move slowly, because the wolves
> need their space; put out your hands so the wolves can smell you. The wolves
> haven't had anything to eat. The group enters the fenced area where the
> wolves live. Mr. Vogelsang, her husband, and also caretaker of the wolves,
> passes the cage on the outside with a bratwurst in his hands and the wolves
> follow him on the other side of the fence, whimpering, no longer paying any
> attention to the group. That is the main difference to dogs, Mrs. Vogelsang
> explains. Wolves are more independent. Without food, there is nothing to gain.
> Apparently, one guy in the group has some treats in his pockets. The wolves
> stick around him, allow him to pet them.
>
> (Field notes and interview transcript, Wolfstag Springe,
> 20 February 2016; translated by the author)

On 20 February 2016, my husband and I visited the 'wildlife' park close to
the village where we live (Wisentgehege Springe 2018). The park focuses on
wisents (European bison) but is also home to wolves. The whole day was
dedicated to wolves: talks about the different wolves in the enclosures by the
couple who takes care of them, watching the wolves being fed, an auction of
a scarf made of wolf wool for a good cause and, at the end, meeting the
wolves in their enclosure under the supervision of their caretakers. The day
was filled with a special atmosphere: not only because of the possibility of
getting close to those animals that fill our minds with various expectations
and stories, but also because of the – at that time – very recent and lively
discussions about one particular free-living wolf who got too close to humans
and how to manage that certain individual with the official name of MT6.[1]
Wolves have been getting close(r). While that day with the wolves allowed and
encouraged a meeting between humans and wolves with almost no distance
between them, it focused, at the same time, on the debates about how to deal
with their free-living counterparts and what space they are allowed to have in
what is called our cultural landscape. How close can they get? What distance
do they need to keep? What happens if they get too close?

That night was the closest I had been to a wolf, until I met that certain MT6 in a museum. A taxidermic statement for the negotiations about how close humans and wolves can(not), respectively, (do not) want to live together. After attempts to scare him off and lengthy debates in the political and public arena, he was legally shot on the evening of 27 April 2016 because of various encounters with humans. Those showed that he was lacking the shyness expected of him and was said to be a possible danger to humans. He was then placed in a taxidermist's hands and, subsequently, put into an exhibition at the Landesmuseum Hannover about the wolf returning to Lower Saxony, a northern state in Germany. This is his story and, simultaneously, an overview of current negotiations between humans with, over or about wolves while they are getting close(r).

Who was MT6/Kurti?

This chapter[2] is about a wolf who was named MT6 by scientists and Kurti by his fans (Zips 2016).[3] The first part deals with his biography, or rather the construction of a non-human[4] biography by humans and how that can be interpreted. It examines the reconstruction of his life – who is participating in telling his story and why – and tries to consider theories about human bio-graphical narratives regarding non-humans. The first part ends with the topic of individualization, and then another perspective is introduced that strengthens the position of the wolf regarding his[5] individuality.

The second part, therefore, analyses the various forms of agency, outlined by Mieke Roscher (2016), that he (the wolf) performed during his lifetime and beyond. It looks at human–wolf relationships and the process of *becoming with* (Haraway 2008, 19), instead of taking for granted given or pre-existing entities. It aims at deepening the understanding of the complex processes in current wolf management endeavours and the various forms of interaction and connections between wolves, humans and other beings. Or as social anthropologist Garry Marvin states in summary:

> At the centre of human–wolf conflicts is not simply a wild animal that might, or might not, be allowed to live its own life on its own terms. Rather, it is *wolf*, a creature that must continue to carry the weight of its cultural creation. This wolf is a creature not just of flesh and blood and of its own habits but also a creature of human moral, social, economic, political, aesthetic and emotional concerns and projections.
>
> (Marvin 2015, 180; emphasis in the original)

In both parts of my analysis, on the one hand, I use material from various visits into different fields in my research area of Lower Saxony, on the other hand, I build on media texts from the last 2 years. My leading question is: What can Kurti's story tell us about wolf management in Lower Saxony? Wolf management is to be understood very broadly. It includes the institutionalized parts of the assemblage that evolves around the return of the wolf to Lower Saxony, consisting mainly of monitoring, public relations and knowledge transfer, documentation, damage prevention and compensation (Wolfsmonitoring 2018). Wolf management cannot, however, be understood only as some fixed institution of different organizations. Even though it is supposed to implement rules of controlling and managing a fixed set of behaviours, it should be conceptualized more as a process of learning from and adapting to various situations, locations and contexts.[6] This conceptualization is based on the assumption that more actors play a role than might first be anticipated; an actor is defined here according to European ethnologist Maria Schwertl referring to sociologist and specialist in Science and Technology Studies, Adele Clarke, as "everything that makes a difference in a situation" (Schwertl 2013, 112; translated by the author). Several other agents emerged with the wolves: such as the wolf-lovers and wolf-haters, the institutions, ministries, hunters, shepherds, horse owners, scientists, sheep, goats, horses, deer, forests, mothers, children, laws, fences, wolf-experts, wolf-ambassadors, police officers, criminal technicians, labs, GPS transmitters, NGOs, guns, cars, cell phones and journalists.

By theorizing wolf management as a process, we get the chance to be more open to new situations, to learn, and to change and to solve conflicts rather than concentrating on failures in a somehow given, unchangeable setting. A broad approach also provides the opportunity to account for the various actors in different fields that enter into relationships that are constantly changing. The analysis of MT6's case regarding, firstly, the constitution of his biography and, thus, his individualization and, secondly, applying different concepts of agency, aims at underlining and understanding this conceptualization of wolf management and tries to open up perspectives.

Before turning to the case of this one specific wolf, I give an insight into the situation in Lower Saxony, a northern state in Germany. Wolves entered Lower Saxony from the east, finding new homes and territories. The first signs of wolf presence in Lower Saxony were found in 2006 (Nds. MU 2017c). It took 6 more years to have the first freeborn puppies in 2012. The latest information in February 2018 (relating to 2016/17; Reding 2018)[7] shows that Lower Saxony is home to 14 packs or families,[8] two couples and three single wolves, which have been confirmed. Eight more areas seem to be inhabited by wolves, but the final proof is still missing. Since 2003, 36 wolves have been found dead in Lower Saxony. Most of them were killed in traffic accidents (NLWKN 2019), while only one dead wolf is marked LE ("*letal entnommen*" – lethally removed, i.e. shot legally). It is this one individual wolf that dictates the discourse in Lower Saxony (if not in Germany). How can the special and outstanding position of this one specific wolf be explained? What does it show about the political and socio-cultural processes that are at play?[9]

Biography of MT6[10]

As social scientists Emmanuel Gouabault and his colleagues point out in their study *On the Personification of Animal Figures in the News*, "[t]he placement of individual figures at the center of media representations increases the potential for emotional connection and proximity and makes the treatment of general interest subjects more concrete, more available, but also more problematic" (Gouabault, Dubied, and Burton-Jeangros 2011, 78). One of the major problems becomes evident when turning to the genre of biography. Non-human animal biographies are not a new phenomenon and have their advantages because of their narrative character and the possibility of showing non-human animals as actors (Krüger, Steinbrecher, and Wischermann 2014, 23). However, following the biography researcher, Winfried Marotzki, who states that "life-history shows itself to be a construct produced by the subject and which, as a unit, organizes the wealth of experiences and events in the course of a life into some coherence" (Marotzki 2004, 103), the question is: Who is the subject in the constitution of Kurti's biography? It is obviously not him telling me in a biographical-narrative interview. Therefore, if humans are to create his story, what are the reasons and the mechanisms behind it? Are humans trying to use his biography for the same reasons they narrate their own biographies, to be specific: "meaning-production, [...] the creation of self-images and world-images [...] meaning-ordering, sense-creating", as summarized by Marotzki (2004, 102f.)? My steps in the analysis are loosely based on Gabriele Rosenthal's analysis of narrative-biographical interviews. I transfer parts of her method to the written statements that constitute Kurti's life history. Her following remarks are of special interest to me: "[...] what events have been enhanced in the narration, what events cannot, or only with difficulty, be made subjects of interest and what kind of sequencing that is at variance with chronology may be set up" (Rosenthal and Fischer-Rosenthal 2004, 263).

The starting point of my analysis is the text that accompanies the exhibit of Kurti in the museum. It only mentions two dates: born in 2014, the fact that he got attention because of not being shy enough and approaching humans and being killed in 2016. For more information about "his fate", the visitor needs to go to another part of the exhibition, a wall with the title "Wolf management" where one gets information on a touchscreen on the general institutions involved in wolf management in Lower Saxony. Moreover, one can follow the lives of three different wolves that rose to fame, one of them being MT6/Kurti. Five slides tell the story, emphasising a technical, biological perspective, filling the holes the other text left and, thus, contrasting the personal encounter between the visitor and Kurti. Who tells this story? The Wolfsbüro (Bureau for the Wolf) and Naturschutzinformation des Niedersächsischen Landesbetriebs für Wasserwirtschaft, Küsten- und Naturschutz (Lower Saxony Water Management, Coastal Defence and Nature Conservation Agency)[11] both departments of the Ministry of the Environment of Lower Saxony, are responsible for the contents of the exhibition.[12] Taking into account that this was most probably written in retrospect, after the death and the very tense environment concerning the first legal shooting of a wolf by the authority's order,[13] this biographical-technical narrative serves as an explanation, leaving out anything that might not be factual. It gives an overview of the actions that were taken to deal with MT6[14] and ends with possible explanations for his behaviour that might be due to human failure. It focuses basically on events that touched the institutionalized parts of wolf management and leaves out events that might be more important from MT6's perspective, probably because it would go from facts to assumptions and, therefore, not fit into the exhibition's overall tone. The narrative, furthermore, encourages prevention, so that a case like MT6 does not occur again. It is in the very last sentence that hope is expressed and a more pro-wolf attitude appears.[15] There are, however, a few more sentences that break the impression of a strictly matter-of-fact text, giving way to an attempt to get inside MT6's mind. One example to clarify my point here: After putting a radio-collar around his neck, the encounters stopped for a little while.[16] However, he then became braver again and appeared close to and interested in cars, dogs and humans. The text states: "He didn't seem to be impressed by the presence of humans".[17] This formulation is not as neutral as the rest of the text and it shows the defining quality of MT6: his curiosity, and this makes him visible as a real sentient being whose thoughts and actions we try to understand.

To come back to Marotzki's processes behind biographization; all of them are to be found in this example. Through the reconstruction of MT6's biography, we engage in meaning-production and meaning-ordering, respectively, sense-creating. Moreover, we create self-images and world-images. Kurti's life story is recreated to show the exception. The rules that are implied in his story reflect how we want to see ourselves when dealing with the wolves, where our and their place in this world is, in short: how to deal with wolves in the right way. In the context given above, humans writing a non-human biography use this narrative not to mirror their (the wolves') meaning-production, meaning-ordering, sense-creating, self- and world-images, but to strengthen and reaffirm ours (the humans').

To conclude, I briefly want to mention another function of biographization: individualization or personification. "Narrating the histories of particular non-humans imparts a sense of their individual lives", writes animal advocate Joan Dunayer (2016, 95). While the focus to individualize wolves lies on a genetic analysis of the institutionalized parts of wolf management,[18] the broader public personalizes through narrating stories, which may be biographic, fairy tales or other genres. What is lacking is the inclusion of the actual being. All those narratives are representations of individuals who treat non-humans as persons on different levels.[19] The underlying general theoretical and methodological questions are yet to be answered. The genre of non-human animal biography enables humans to connect with entities around us; it shows how non-human and human lives are interwoven and can help to make something we do not seem to understand more understandable.[20] In the following part, therefore, I want to emphasise the wolf's perspective and the interconnections between different parts of the network that constitute wolf management in the broad understanding of that term outlined above.

Agencies of MT6

Mieke Roscher (2016) outlines different forms of agency that I want to use for analysing the impact of the wolf MT6/Kurti in different situations and contexts. Agency unfolds itself and has to be thought of in relationships, spaces and times. Let me elaborate this briefly: Shared spaces (Kurth, Dornenzweig, and Wirth 2016, 15), like streets (wolves use human-built infrastructures), can be a danger for all participants. The chart (NLWKN 2019) only shows wolfish victims, but who knows what happened to the drivers or to the cars. In this particular situation, they build a network in which each action of each participant has different consequences for all the individual parts. The effects of this specific agency reach beyond the defined space and time: One could think of discussions about bridges over highways or the most dangerous streets (cf. NABU 2014, 2018), or even a general doubt whether the wolf can survive in a territory like Lower Saxony.

Roscher uses the term "relational agency" to describe "face-to-face interaction between humans and animals, respectively, a single human and a single animal" (Roscher 2016, 57; translated by the author). She refers to the network between actors and their environments with the term "entangled agency" (Roscher 2016, 58). "Embodied agency" refers to the corporeality of animals and the materialities of actions in certain shared spaces that also involve conflicts (Roscher 2016, 59). The last form is the specific agency of certain species called "animal agency" (Roscher 2016, 60).

"Relational agency": meeting wolf, building relationships

> I almost miss the entrance to the wolf exhibition that is part of the larger Nature Worlds in the Landesmuseum Hannover. It is a small L-shaped room and I go around and discover Kurti, almost hidden in a corner. A spotlight seems to imitate the sunlight through the leaves of a forest. He looks towards the visitor, standing on a stone or stump, a patch of grass and leaves around him. Depending on where I position myself, trying to get a closer look at his face, he seems to be cross-eyed or watching me with his left eyebrow raised. One can find a short biography of this particular wolf on one of the birches in the background of the diorama. I am eye to eye with a wolf – a stuffed one.
>
> (Field notes, exhibition Landesmuseum Hannover, 21 May 2017;
> translated by the author)

He has an effect on me: I feel sorry; I am astonished; I wonder what his fur must feel like. There is something that he does to me, even though (or maybe, because) he is dead.[21] He affects me because I realise that he is vulnerable, endangered and mortal – just like me (Huth 2014, 63). His corporeality builds a bridge between us (Balgar 2016, 137), thus, making us both subjects in a relationship (Ohrem 2016, 70) in this certain situation in this particular shared space.

It is in those shared spaces where *relational agency* unfolds itself, where and when relationships are built, between the wolf, the human and the expectations of both. While we cannot be sure about the wolf's expectations concerning its counterpart, there are various backgrounds that influence the situation on the human side and need to be considered when describing and analysing a situation where wolf and human meet. Those backgrounds could be scientific knowledge, lay knowledge, rumours, fairy tales or other narratives. All have an impact of some sort and, therefore, should not be judged or valued but described and analysed. The situations of meeting MT6 were rare but had a big impact on those who encountered him. And they are not limited to the living wolf. Due to his presence in the exhibition at the Landesmuseum Hannover, visitors finally get the chance to meet MT6. A very special relationship between wolf and human is described by the taxidermist who worked on Kurti for hundreds of hours. For him, the main reason to do his work lies in his mission to "bring animals closer to the people, and as naturally as possible" (Kressel 2017a; translated by the author). Thus, for the taxidermist, it is the goal to recreate an exact copy of Kurti, the individual wolf, not just a representation. A Swiss taxidermist who was interviewed for an exhibition in Bern, Switzerland, described her work and emphasized the importance of having a wolf as an exhibit in similar terms of creating a relationship: "The exhibit gives people the opportunity to meet the animal themselves. They can stand in front of the animal, look into its eyes and deal with it" (Alpines Museum der Schweiz and Universität Zürich – ISEK 2017, 28; translated by the author).

According to social scientists and human–animal studies researchers, Markus Kurth, Katharina Dornenzweig and Sven Wirth, agency can be framed in the fields of "intentionality of action", "possibility/capacity of action, respectively, performance of action" and "effects of action" (Kurth et al. 2016, 16; translated by the author). While encounters with MT6 during his lifetime were structured mainly by the possibility of his actions and agency, visitors in the museum speak about and show signs of certain effects of his presence that somehow contradict expectations built up in his lifetime. Some examples to illustrate my point here: Media coverage of encounters with MT6 was broad. The descriptions of meeting MT6 focus on the fear of 'what could happen', a possible action. A woman who encountered MT6 on a walk with her baby is quoted: "He was big, I feared for myself and my child" (Doelke and Berger 2016, 26; translated by the author). The picture of the 'big bad wolf' is invoked and strengthened. Visitors take this impression with them to the museum where it gets contradicted. Seeing the stuffed MT6, I hear a mother telling her toddler: "Look, a woof-woof"[22]. Other visitors comment on MT6's small size. A woman states: "I imagined Kurti being way bigger" (Field notes, exhibition Landesmuseum Hannover, 21 May 2017; translated by the author). Those relationships again differ from the one between the (at that time) Minister for the Environment, Stefan Wenzel (Green Party), and the exhibit: For him, Kurti shows "what we as humans can do wrong" (Benne

2017, 6; translated by the author); he serves as an "admonitory example" (Kressel 2017c). The minister's reluctance to be photographed alone with the exhibit and his use only of the technical name MT6 is a summary of the relationship those two built up in the past years (Benne 2017, 6). The whole story behind the wolf's unusual behaviour resembles an episode of a television crime series where different relationships between humans and MT6 are analysed to find an explanation to prevent a similar case occurring. One and a half months after his death, an article deals with "photographs and films that up until now were a well-kept secret" (Kressel 2016a; translated by the author). Theo Grüntjens, one of the approximately 130 advisors on wolves in Lower Saxony, is quoted: "Even though the wolves themselves didn't behave in the wrong way, they learn from a combination of many close meetings that humans belong to wolfish life, that they are no danger" (Kressel 2016a; translated by the author). Klaus Bullerjahn, also an advisor on wolves, states: "humans had made them into what they in individual cases became – less shy wolves that were made accustomed to humans" (Kressel 2016a; translated by the author). The article mentions photographs and films of wolves that were taken by soldiers, rangers and firefighters when those wolves were still puppies (under the age of 1 year). All of them were siblings of Kurti or part of his family. Getting close(r) is, thus, not only describing a one-way path: wolves approaching humans and human territory. It is implying and trying to grasp the mutual movement: humans and wolves approaching each other interdependently.

From relationship to network:[23] entangled agency

At the same time, those examples show another form of agency that points out the network character. Single face-to-face interactions and relationships are to be seen as entangled in bigger frameworks; Roscher uses the term "entangled agency" (Roscher 2016, 58). The outcome of a situation might be taken into the media and used by various other actors who had not been involved directly in the happening before but are part of what can be theorized as "entangled agency": that means, the network is more than just a localized situation. It has its place in history and may change it. It goes beyond the local and must be seen within a large framework as well. This term also points to the environments and nature-cultures (Haraway 2005, 12, 16) in which networks are situated and of which networks are made. I deal with two different networks in the following, that are separated here for analytical purposes, but not in the daily lives of the participants: firstly, I want to include technical artefacts in my analysis that have been a bit out of the focus so far; secondly, I consider Kurti's effects on the political realm.

Wolf researchers focus not only on genetic analysis but also on telemetry to track wolf presence. So far, only two wolves have been fitted with a radio collar in Lower Saxony: MT6 and his sister, FT10. The signs of this non-human technological network are still visible on the exhibit: "The visible abrasions on the neck are marks from the radio-collar" (Kressel 2017c; translated by the author). By tracking his movements, researchers hoped to get more insights into his life and thus justify (if necessary, legally) the information and recommendations they gave to the authorities in how to deal with him in various situations. A problem occurred on 20 December 2015 when the GPS stopped functioning. Telemetry, of course, only records their movements; other measures are used to try and protect against them. Fences are one example. To get financial aid in case of a wolf attack on livestock, owners are obliged to make sure they have done everything possible for the safety of their livestock (NLWKN 2017, 5). As the result of protests by livestock owners and other interest groups, with permission from the EU Commission the financial aid was raised from 15,000 Euro in three years to up to 30,000 Euro for each year for every business (Nds. MU 2017a). However, livestock owners still have to prove that the attack was by a wolf and the procedure is not easy and takes a while. One case as an example is the following: an article reports the ending of a case on 18 May 2017. A wolf was thought to have bitten through an electric fence and killed six young animals, kids of goats and lambs of *Heidschnucke* (German grey heath) sheep two months earlier. While nobody could explain the hole in the fence and a wolf was not found to be responsible, genetic analysis eventually proved that a wolf had killed the animals (NDR 2017).

What can be seen here are networks that blur the boundaries between the natural and the cultural; entities become entangled in something new. Morten Tønnessen points out the paradox that is inherent in these findings: "The future of the wolf, a master of seclusion, apparently depends on its being managed by conservationists to an extent that makes the very notion of 'wild wolves' appear dubious" (Tønnessen 2010, 1). Categories that were sharp once, need to be negotiated anew. In the entangled worlds in which we live. Kurti has not only mobilized this discussion, but he has also set a precedent (Kressel 2017b)[24] for dealing with wolves from a political perspective.

His case was a measure by which the officials could retrieve the public's trust in the official wolf management, i.e. if there is a problem, the officials will not leave you alone, but will deal with it. But it did not work out. Kurti, once taken into the political arena, has become a symbol of the government's failure to deal with wolves and is seen as proof of non-functioning wolf management.[25] While the political elite does not seem to be able to handle the problems correctly, other groups are taking over. One example is a communal fraction of the CDU (Christian Democratic Union) party who decided to resettle a forest kindergarten. Since they did not include the groups affected (e.g. caretakers, parents, children), it was not put into action. While the members of the CDU thought that children, caretakers and parents were scared of being in the forest, the supposed fear was non-existent due to the explanations given by the local advisor on wolves (Kressel 2017b). Other groups have reasons to be scared though and organise various forms of protest against wolves: Some livestock owners brought dead sheep to the parliament and since May 2017, there have been bonfires lit all over Germany (and Europe) to make their voices heard (Kressel 2017b). The wolf has a prominent position in the current election battle for the new parliament in Lower Saxony.[26]

"Wolves don't eat grass": embodied agency

Effects and consequences, such as changing attitudes towards wolves because of a personal encounter, also need to be viewed from a corporeal perspective, especially in the case of MT6. The tipping point in his story are the encounters with humans who judged him as being too close, too unwolfish (i.e. not shy enough). He not only approached and followed a woman with a baby buggy but was also seen in front of a refugee camp, where he stayed overnight and was suspected of having attacked a dog and followed two young boys from the forest into their village (Gude 2016, and various other media reports). This materiality of the wolf's body can be theorized as a performance of "embodied agency" (Roscher 2016, 59) that acts upon the vulnerability (Ohrem 2016) of other bodies. While I have focused so far mainly on human–wolf relationships, it is now that other-than-human bodies play a major role in the discourses, especially those of sheep, cows and horses but not only those. As mentioned briefly above, from May 2017 onwards, associations of sheep breeders, herders and farmers have organized bonfires to get publicity for their cause. I visited one of them on 12 May that had the title: "Wolves do not eat grass" (WNON 2017).[27] Talking to the organiser of this bonfire, he explained his point of view that summarises my findings from other field visits (Field notes, bonfire, 12 May 2017, 2f.). It is not a fundamental rejection of wolves that he expresses. Rather, he longs to be taken seriously concerning his fears and his work. He criticises the devaluation of his occupation and rural life in general by those putting the wolf on a pedestal, untouchable and higher than all other lifeforms. These findings are comparable to the results of studies elsewhere in Europe that I will introduce briefly. Socio-logical research dealing with the return of wolves in Norway and France have shown that conflicts around the wolf evolve from neglecting lay and local knowledge (Skogen 2001, 218–220). Therefore, they can be described and ana-lysed as conflicts between 'urban elites' and 'rural communities'[28]. The wolf becomes an "icon of urbanity" (Skogen and Krange 2003, 320) which has two dimensions: "a threat imposed upon rural communities by urban elites or […] an object of hegemonic and patronizing academic knowledge" (Skogen, Mauz, and Krange 2008, 106). The conflicts also point to different understandings and uses of so-called nature:

> While farmers and hunters saw the natural physical environment as a landscape for sustainable *use*, as productive areas for logging, grazing, hunting and berry picking, the informants who expressed positive views on the presence of wolves saw this same environment as untouched nature, or *wilderness*.
>
> (Figari and Skogen 2011, 327)

As European ethnologist Michaela Fenske argues in her work about urban bee-keeping, the definition of nature and culture in the Anthropocene is one key question in late modernity's societies:

Are Culture and Nature separated spheres? Is Nature the untouched wild-ness that should be left alone and be protected? Or is Nature hybrid, something that has become/emerged socio-ecologically, that is constructed by and through human use? *[Sind Kultur und Natur getrennte Sphären? Ist Natur die unberührte Wildheit, die es als solche sich selbst zu überlassen und zu schützen gilt? Oder ist Natur hybrid, ein sozial-ökologisch Gewordenes, das gerade auch durch menschliche Nutzung gestaltet wird?]*

(Fenske 2017, 37; translated by the author)

Moreover, the debates show a battle for participation and against hierarchies:

The network members talk about an antagonistic relationship between powerful circles with an urban basis and a powerless group living in rural areas. The urban–rural dichotomy is experienced as a deep and many-faceted conflict. At the core is an uneven relation of power.[29]

(Krange and Skogen 2011, 477)

The organiser of the bonfire I talked to also strengthened the point of the valuation of nature and species protection. He spoke about the sensible ecosys-tem that shepherds are taking care of in the Lüneburger Heide, an area in the northeast of Lower Saxony. He told me that nature would be reduced if not for their work. He also mentioned his work for the local birds and how he adjusted his routines to make sure they would survive in otherwise heavily farmed land-scapes. Moreover, he is active in campaigning against a nearby city that takes out too much water from the area with the result that trees die. At the same time, he questions the motives of groups who support the wolf, because for him, too many wolves will destroy the quality of nearby touristic places that city people use as retreats and "to fly to Gran Canaria for a vacation is not really ecological either" (Field notes, bonfire, 12 May 2017, 3; translated by the author). Con-cerning human bodies, he talks about specific groups that are labelled as espe-cially vulnerable: children, women and the elderly. This is also the case for a small community in Norway, the inhabitants of which were interviewed by environmental sociologists Ketil Skogen and Olve Krange:

It is frequently claimed that a principal asset of life in rural areas, namely outdoor recreation, is seriously devalued because many people are afraid to go for walks and especially to take their dogs out. There is a particular concern for small children who are allegedly not allowed to play outdoors alone anymore. Elderly people, especially women, are seen as another strongly affected group. By picturing 'weak groups' as vulnerable to dangers that are imposed upon the community from outside, they emphasise that wolf protection is cruel and inhuman. It represents an infringement on the community, and is in effect an assault on the 'weakest among us'.

(Skogen and Krange 2003, 317)

As those comments show, performance of embodied agency evokes a corporeal response that is then shifted into a cultural system called emotions. Regarding MT6, those responses were very strong and mostly negative. The effect of this situational relationship was fear that was transferred into language and narratives which, on the one hand, have their roots in cultural and collective memory (such as Little Red Riding Hood, of course) and, on the other hand, build upon developing new scenarios. I came across one example to clarify my point during a public discussion in the city of Hildesheim in Lower Saxony. It shows the development of a worst-case scenario as it happened between various actors on stage and in the audience. The story began with the case of more than one wolf breaking into a flock of sheep, which was later corrected to have happened with dogs and not wolves (Field notes, public discussion Hildesheim, 5 June 2017, 4). The story went on: What if a wolf broke into a herd of sheep, making them panic and flee their territory, running onto a street and dying in traffic accidents? Who would pay for the damage? And even further: What if such an accident led to a person being left paraplegic? (Field notes, public discussion Hildesheim, 5 June 2017, 6 and 8). The organiser of the bonfire made a similar point by stating that the current happenings to "just farm animals" did not produce enough attention. He thinks a killed dog or child would turn around the discussion and lead to the measures he considers necessary.

While those findings focus on keeping wolves out of human life-worlds or environments, minimizing contact and the effects of embodied agency, other practices try to enable encounters and create 'contact zones' (Haraway 2008), since there is also a longing to meet the 'other', a curiosity on both sides. One way to do the latter are wildlife parks and centres for wolves, another is the exhibition of MT6. Even in his death, he performs embodied agency. The taxidermist, therefore, chose "a standing, frozen-in-movement model. A bit of body tension should be displayed" (Kressel 2017a; translated by the author). He wanted to create "the perfect body" (Kressel 2017a; translated by the author) and gave him a lifelike appearance by painting his nose so that it "seems moist and glossy" (Kressel 2017c; translated by the author). At the same time, the exhibit of Kurti in the museum was heavily criticised by PETA and other animal protectionists who see it as "picking over the bones" and that it is "impious" to show dead animals in museums. The life of wolves should be shown and experienced elsewhere (Benne 2017, 6; translated by the author). Another way of creating contact zones besides zoos or museums, is the promotion of wolf tourism. In Lower Saxony, one can find professionally organized and accompanied tours for lay scientists that aim to support the institutionalized wolf management. People can buy one week of following wolf tracks and collecting samples for 1770 Euro. The expeditions are criticised by the Hunters' Association of Lower Saxony (LJN) which is officially responsible for wolf-monitoring because they are perceived to devalue the work of locally engaged volunteers by making huge profits with "sensation tourists", disturbing local fauna and flora and breaking the contract between the LJN and NLWKN (PAZ 2017; Randt 2017; translated by the author).

Another offer that is not as controversial is a bike tour through wolf-land with information concerning the return of wolves (Reiseland Niedersachsen 2017). Both might never result in an actual encounter, but they address the longing for it that is a result of various forms of wolfish agency as outlined above. It is in those contact zones that getting close(r) is practised and negotiated.

"Animal agency": (un-)wolfishness

Taking the difference between 'the' wolf and 'wolves' seriously, my last point focuses on what Mieke Roscher calls "animal agency" (Roscher 2016, 60). It refers to the specific agency of different species and their own being in the world that leads to unexpected situations with other actors. While other so-called wild animals have returned to Lower Saxony almost unrecognized, the extraordinary status of wolves might be explained by referring to their specific agency. The case of MT6 clearly shows the difference between just behaving and performing "animal agency". Comments such as: "He is not at all behaving like he should" (Doelke and Berger 2016, 26; translated by the author) or "But the wolf does not cooperate" (Gude 2016; translated by the author) imply expectations concerning the behaviour of wolves in general and how the wolf as an agent performs against those. The catalogue of what is typical wolfish behaviour and what is not and how untypical behaviour such as Kurti's might be explained has already been dealt with above. However, it is of importance again, since wolves as a species are under special watch. All (found) dead wolves are given to the Leibniz-Institut für Zoo- und Wild-tierforschung in Berlin, where they are checked for possible diseases and the causes of death are determined. The procedures are similar to human crime investigations and the technical, financial and temporal efforts are enormous. The examination of Kurti resulted in "no evidence of health problems" and, therefore, the focus shifted to human influences (Kressel 2017d; translated by the author). Anticipated roles are then exchanged: While the wolf is not acting and behaving as he should, but performing his own choices, the humans are told how to behave; a brochure with advice for the correct behaviour in an encounter with a wolf (NLWKN 2016) can be interpreted as a tool that tries to give back to human agents the control that has been taken away by the wolf. It also emphasises a process of learning how to deal with this new neighbour.

Getting close(r), while keeping distance

I began this chapter with the title "Getting close(r)", telling the story of Bene who was allowed to enter the wolves' area at a local wildlife centre. The dates that offer a direct contact are always fully booked in advance and more offers are being created.[30] There seems to be a longing to meet wolves. Similarly, in the exhibition at the Landesmuseum Hannover: I visited Kurti again on 13

October 2017 before he was taken to another museum in the Lüneburger Heide. This time he was behind glass. I asked the person at the cashier desk why this measure had been taken. She answered that too many people touched him and even allowed their children to climb onto him for a photo. He needed protection – again.

At the same time, free-living wolves are coming closer to us. They enter areas that were formerly firmly in human and other non-human 'hands'. Lower Saxony and other states of Germany to which wolves are currently returning are in the middle of a process. Plans have been made how to deal with wolves once they enter a new territory, but they seem to be failing. Wolf management, in the usual understanding of the term, suggests a perfect plan, a rulebook by which humans can act and that wolves follow. As shown in this article, that is a misleading concept that strengthens conflict, opposition and frustration. It also implies a top-down policy and instrument of power of some but not all interest groups that, according to my respondents, is exclusive and does not consider their perspectives. What can be stated is that wolf management is part of a broader process in society that has been described as a shift from a disciplining to a controlling society (Kurth 2016, 190). The analysis of Kurti's biography showed how, in retrospect, this narrative is used to explain failures in controlling an individual non-human animal and to strengthen our views on wolves. Theoretically, the life history of MT6 can be framed as moments of *transdifference*. This term "describes situations in which obsolete constructions of difference based on a logic of binary oppositions become fluid and temporarily lose their validity without being fully and finally deconstructed [*bezeichnet Situationen, in denen die überkommenen Differenzkonstruktionen auf der Basis einer binären Ordnungslogik gleichsam ins Schwimmen geraten und in ihrer Gültigkeit temporär suspendiert werden, ohne dass sie damit endgültig dekonstruiert würden*]" (Lösch 2005, 27 as cited in Kurth 2016, 195; translated by the author). Thus, what is happening in the debates on the return of wolves can be seen as a society dealing with changes, in what Zygmunt Bauman called "liquid modernity" (Bauman 2000). Wolves may serve as an amplifier of changes that are already happening and make our lives more fragile, temporary and vulnerable. Kurti, however, is used as an example of how coexistence is not favoured. Even though he breaks the rules and transgresses boundaries in certain given situations, there is no effect on the general thinking about how we want to live with wolves. From a human perspective, he is transformed into an individual that becomes an icon and a symbol but no longer serves to represent his species. Kurti is the one specific wolf that was too unwolfish during his life to be kept alive. At the same time, in his death, he serves as an ambassador for the remaining wolves and those yet to come, even though humans made him an outsider, an outlaw that could not be tolerated. He is simultaneously the good and the bad wolf and, thus, stands for the ambivalent relationship we have towards and with wolves. I showed how various those relationships are in the second part by using the different agency concepts. I tried to take seriously the call for more studies that consider "the animal itself or the human-animal-relationship" (Krüger et al. 2014, 19; translated by the author) and balance human and other-than-human perspectives in the current wolf management. Factors that play a role in this complex process are shared spaces, relationships and networks, expectations and different forms of knowledge, as well as corporeality and vulnerability. It is not enough to argue with biological and ecological facts; socio-cultural, economic and political contexts must also be considered. Perhaps the entanglements of humans and wolves in Lower Saxony can be described with what Bernhard Tschofen called "complementary species",[31] following and expanding Donna Haraway's idea of "companion species". It grasps the constant and multidimensional movements of humans and wolves between fascination, attraction and rejection because of similarities and differences. It might help us to understand the practices of getting close(r).

Notes

1 M = male, T = telemetry, 6 = he was the sixth wolf to get fitted with a radio collar in Germany.
2 This chapter is based on my presentation at the EASA conference in Milan in July 2016. It takes in further developments and recent events. I am thankful to Michaela Fenske and Bernhard Tschofen for not only organizing the panel and giving me the opportunity to talk there, but also for helping me to transform a 20-minute talk into a paper for a publication. Thanks to Philip Saunders and Alison Jones for proofreading, and to Elisa Frank, Nikolaus Heinzer and Marlis Heyer for constant support. I wish to thank my husband Bene for accompanying me in this process with unconditional love, warmth and thoughts.
3 See also Dunayer (2016, 94): "Identifying nonhuman animals by their personal names emphasizes their individuality".
4 The term non-human has been criticised for still using human as a point of reference. Since other terms also have weaknesses (human and other animals, other-than-human: all are based on the dichotomy even by trying to break it), I stick with non-human in this paper. Cf. Wischermann (2014, 103).
5 I use gendered pronouns since this chapter is dealing with a highly individualized non-human animal, see Dunayer (2016, 94): "The gendered pronouns *she* and *he* [emphases in the original, I.A.] help to individualize non-human animals".
6 This conceptualization and understanding of wolf management is one of the key points in the research design of the project "Die Rückkehr der Wölfe. Kulturanthropologische Studien zum Prozess des Wolfsmanagements in der Bundesrepublik Deutschland" ("The return of wolves. Cultural anthropological studies concerning the process of wolf management in the Federal Republic of Germany"; translated by the author), led by Prof. Dr. Michaela Fenske, funded by Deutsche Forschungsgemeinschaft (DFG). My PhD project is one of two pieces of research being done in this project.
7 It is not that easy to give an absolute number of wolves due to mobility, death and other causes. The official monitoring states 84 confirmed wolves for the year 2016/2017, admitting that "the real number of individuals could be considerably higher" (Reding 2018, 19; translated by the author).
8 I refer to family instead of pack because most of the newest research suggests such a reading of the group structure of wolves.

9 For studies in other European countries that deal with similar questions, cf. Campion-Vincent (2005), Lescureux (2006), Skogen (2001), Skogen and Krange (2003), Skogen, Mauz, and Krange (2008) and Figari and Skogen (2011).

10 The concept of animal biography was discussed recently at the conference "Animal Biographies – Recovering Animal Selfhood through Interdisciplinary Narration?", 9–11 March 2016 in Kassel. Accessed 1 November 2017. https://www.uni-kassel.de/fb05/fachgruppen/geschichte/human-animal-studies/konferenzen.html

11 Cf. NLWKN, "Welcome to NLWKN".

12 Cf. NLWKN: "Das Wolfsbüro des NLWKN".

13 A total of 150 charges were filed against the Minister of the Environment as he pointed out again in a speech in the state's parliament on the 20 September 2017: "Und wir haben – das bitte ich nicht zu vergessen – mit dem Wolf MT 6 erstmals in Deutschland einen Wolf erschießen lassen. Diese Maßnahme war seinerzeit notwendig, sie war schwierig durchzuführen und sie hat mir persönlich 150 Strafanzeigen eingebracht" (Nds. MU 2017b). [And we had – I ask you not to forget that – with the case of the wolf MT6 for the first time in Germany we had a wolf shot. That measure was necessary at that time, it was difficult to execute and it brought 150 charges against me personally; translated by the author].

14 Cf. Photos, exhibition Landesmuseum Hannover, 21 May 2017: 20170521_143307 entitled "Besenderung" (telemetry), 20170521_143424 and 20170521_143517 entitled "Vergrämung" (scare (off)) and 20170521_143535 entitled "Tötung" (killing).

15 Cf. Photo, exhibition Landesmuseum Hannover, 21 May 2017: 20170521_143613 entitled "Ursachen" (reasons/causes): "Denn der Umgang des Menschen mit dem Wolf wird darüber entscheiden, ob diese Tierart hier langfristig heimisch werden kann und das Schicksal von MT6 ein Ausnahmefall bleibt" (The handling of the wolf by humans will decide if this species can become part of the local environment again and if the fate of MT6 will be an exception; translated by the author).

16 Scandinavian researchers support this finding, cf. Arnemo and Fahlman (2007, 6): "Animals that have been captured before (especially wolves) will usually run for cover when they hear the helicopter".

17 Photo, exhibition Landesmuseum Hannover, 21 May 2017: 20170521_143424: "Dabei ließ er sich durch die Anwesenheit von Menschen nur wenig beeindrucken" (translated by the author).

18 Cf. the Swiss collection on the topic: KORA, "Wolf Genetics". And the German centre for "Naturschutzgenetik" (nature protection genetics) that is part of the Senckenberg research institute: Senckenberg Forschungsinstitut und Naturmuseum Frankfurt (2017) "Naturschutzgenetik".

19 A more detailed analysis on the personification process as outlined by Gouabault, Dubied, and Burton-Jeangros (2011) and carried out using the example of the polar bear Knut is also of high interest here but needs to be done in another paper.

20 Cf. Krüger, Steinbrecher, and Wischermann (2014, 23): "Solche Tierbiographien [sic!] erzählen Geschichten von Tieren als Kulturwesen, also von tierlichem Leben, welches eng mit menschlichem verwoben ist. […] Entsteht eine tierliche Biografie in jedem Falle erst im Zusammenhang mit menschlicher Autorschaft und einem menschlichen *companion*?" ["Such animal biographies tell stories about animals as cultural beings, about animal life that is entangled with human life. […] Is an animal biography only created in connection with human authorship and a human companion?"; emphasis in the original; translated by the author].

21 Various perspectives dealing with 'Animals and Death' are to be found in Ullrich and Ulrich (2014).

22 Field notes, exhibition Landesmuseum Hannover, 21 May 2017, 2: "Guck mal, hier ist ein Wau-Wau". 'Wau-Wau' is a toddler-term for dog, using the barking noise as a replacement for the actual word.

23 I use the term 'network' following different theoretical paths. It is to be understood more as a working tool that helps me at the moment to work with and through my materials while the exact theoretical fine-tuning still needs to be done in the process of my research. I take inspiration from Bruno Latour's (2010) *ANT* as well as Jane Bennett's *agentic assemblages* (2010).

24 In the 'Joint statement of the Federal Minister for the Environment Barbara Hendricks and Prime Minister of Lower Saxony Stephan Weil concerning the conflict between wolf protection and grazing land husbandry' Kurti/MT6 in his role as a precedent is brought up again: "Die Sicherheit der Menschen hat oberste Priorität. Wölfe, die sich Menschen gegenüber auffällig verhalten, sind zu beobachten und gegebenenfalls zu töten. Eine entsprechende Entnahme wurde in Niedersachsen bereits einmal vorgenommen (Tötung des Wolfes MT6)" [The safety of humans is the first priority. Wolves who behave peculiarly in the face of humans are to be watched and, if necessary, killed. There has been one such removal in Lower Saxony (killing of the wolf MT6)] (Hendricks and Weil 2017; translated by the author).

25 One example is the foundation of the 'alliance active wolf management' that demands a 'real' wolf management. Cf. Aktionsbündnis aktives Wolfsmanagement, "I <3 Weidetiere." Accessed 1 November 2017. http://www.aktives-wolfsmanagement.de/.

26 Unfortunately, as this is ongoing, I cannot go deeper into the matter in the current article, but it does deserve to be written and talked about in more depth elsewhere. Shortly before finishing the work on this article, the following changes that were promised during the election campaigns are being implemented: improvement of the financial aid (cf. Nds. MU 2017d) as well as equipping more wolves with radio collars (cf. NDR 2018a, 2018b).

27 It is also the official slogan of the organization of livestock owners in the northeast of Lower Saxony (WNON).

28 I put those group terms in quotation marks since they do not accurately reflect the complexity of the situation but are rather a term out of the fields of study (Skogen and Krange 2003, 310).

29 And in similar tone for France see Campion-Vincent (2005, 115f.): "For local human populations that have to cope with the disturbances caused by wolves' presence, the return of wolves corresponds to the intrusion of city dwellers in their narrowing universe. They feel these city dwellers dictate the rules of management of their environment and that they almost live in 'third type zoos' organized by naturalists".

30 Cf., for example, offers by the wildlife park in Wisentgehege Springe (2018) or the Wolfcenter (2018) in Dörverden that can be found on their websites.

31 He used this term based on the insights he got from reading an earlier version of this paper in a lecture he gave at the department of European Ethnology/ Volkskunde at the University of Würzburg on 24 January 2018.

Bibliography

Aktionsbündnis aktives Wolfsmanagement, 2017. I <3 Weidetiere. Available at http://www.aktives-wolfsmanagement.de/ [accessed 1 November 2017].

Alpines Museum der Schweiz (Hächler, B.) and Universität Zürich – ISEK (Tschofen, B.), eds. 2017. *Der Wolf ist da: Eine Menschenausstellung*. Bern: Alpines Museum der Schweiz.

Arnemo, J. M. and Å. Fahlman, eds. 2007. *Biomedical Protocols for Free-ranging Brown Bears, Gray Wolves, Wolverines and Lynx*. Tromsø: Norwegian School of Veterinary Science.

Balgar, K. 2016. "Leiblichkeit und tierliche Agency. Die Handlungsfähigkeit von Tieren im Kontext von Leiblichkeitskonzepten." In *Das Handeln der Tiere. Tierliche Agency im Fokus der Human-Animal Studies*, edited by S. Wirth, A. Laue, M. Kurth, K. Dornenzweig, L. Bossert, and K. Balgar, 137–148. Bielefeld: Transcript.

Bauman, Z. 2000. *Liquid Modernity*. Cambridge: Polity Press.

Benne, S. 2017. "Mit Haut und Haar: Der erlegte 'Problemwolf Kurti' wird zum Museumsstück – und der Minister hofft, dass die Menschen ihre Lehren aus seinem Schicksal ziehen." *Hildesheimer Allgemeine Zeitung*, 20 May 2017, 6.

Bennett, J. 2010. *Vibrant Matter. A Political Ecology of Things*. Durham, NC and London: Duke University Press.

Campion-Vincent, V. 2005. "The Restoration of Wolves in France. Story, Conflicts and Uses of Rumor." In *Mad About Wildlife. Looking at Social Conflict Over Wildlife*, edited by A. Herda-Rapp and T. L. Goedeke, 99–122. Leiden and Boston, MA: Brill.

Doelke, K. and M. B. Berger. 2016. "'Man hat uns den Wolf anders verkauft.' Anwohner fühlen sich in der Wolfsfrage vom Land alleingelassen – und Minister Wenzel ändert seinen Kurs." *Hildesheimer Allgemeine Zeitung*, 19 February 2016, 26.

Dunayer, J. 2016. "Mixed Messages. Opinion Pieces by Representatives of US Non-human-Advocacy Organizations." In *Critical Animal and Media Studies. Communication for Nonhuman Animal Advocacy*, edited by N. Almiron, M. Cole, and C. P. Freeman, 91–106. New York and London: Routledge.

Fenske, M. 2017. "Retten und gerettet werden. Europäische Honigbienen und Menschen im urbanen Resonanzraum." In *Hessische Blätter für Volks- und Kulturforschung*, 35–49. Marburg: Jonas Verlag.

Figari, H. and K. Skogen. 2011. "Social Representations of the Wolf." *Acta Sociologica* 54 (4), 317–332.

Gouabault, E., A. Dubied, and C. Burton-Jeangros. 2011. "Genuine Zoocentrism or Dogged Anthropocentrism? On the Personification of Animal Figures in the News." *Humanimalia* 3 (1), 77–100.

Gude, H. 2016. "Kurti soll sich fürchten." *Der Spiegel*. 12 March 2016. http://www.sp iegel.de/spiegel/print/d-143591178.html [accessed 10 September 2017].

Haraway, D. J. [2003] 2005. *The Companion Species Manifesto: Dogs, People, and Significant Otherness*. 3rd edition. Chicago, IL: University of Chicago Press.

Haraway, D. J. 2008. *When Species Meet*. Posthumanities, Vol. 3. Minneapolis, MN and London: University of Minnesota Press.

Hendricks, B. and S. Weil, 2017. "Weidetierhaltern in Wolfsgebieten helfen – Jagdrecht ist keine Lösung. Gemeinsame Erklärung von Bundesumweltministerin Barbara Hendricks und Ministerpräsident Stephan Weil (Niedersachsen) zum Konflikt zwischen Wolfsschutz und Weidetierhaltung. Niedersächsiche Staatkanzlei." 28 September 2017. https://www.stk.niedersachsen.de/aktuelles/presseinformationen/bund-und-la nd-einigen-sich-auf-gemeinsames-vorgehen-beim-thema-wolf-weidetierhaltern-in-wolfs gebieten-helfen-jagdrecht-ist-keine-loesung-158309.html [accessed 21 February 2018].

Huth, M. 2014. "Ihr Tod geht uns an. Eine Phänomenologie des Sterbens von Tieren." In *Tiere und Tod*, edited by J. Ullrich and A. Ulric, 59–71. Berlin: Neofelis.

Kressel, U. 2016. "Haben Menschen 'Kurti' zum Problem-Wolf gemacht?" *Norddeutscher Rundfunk*. 14 July 2016. http://www.ndr.de/nachrichten/niedersachsen/lue neburg_heide_unterelbe/Haben-Menschen-Kurti-zum-Problem-Wolf-gemacht,woelf e518.html [accessed 30 June 2017]. Only in printed version.

Kressel, U. 2017a. "Wolf 'Kurti' wird fürs Landesmuseum präpariert." *Norddeutscher Rundfunk*. 9 March 2017. http://www.ndr.de/nachrichten/niedersachsen/hannover_

weser-leinegebiet/Wolf-Kurti-wird-fuers-Landesmuseum-praepariert.wolf2986.html [accessed 30 June 2017] Only in printed version.

Kressel, U. 2017b. "Ein Jahr nach 'Kurti's' Tod: Aufruhr im Wolfsland." *Norddeutscher Rundfunk.* 27 April 2017. http://www.ndr.de/nachrichten/niedersachsen/lueneburg_ heide_unterelbe/Ein-Jahr-nach-Kurtis-Tod-Aufruhr-im-Wolfsland,wolf3088.html [accessed 25 March 2018].

Kressel, U. 2017c. "'Kurti' ist jetzt museumsreif." *Norddeutscher Rundfunk.* 19 May 2017. https://www.ndr.de/kultur/Kurti-ist-jetzt-museumsreif,wolf3114.html [accessed 25 March 2018].

Kressel, U. 2017d. "Wo tote Wölfe Arbeitsalltag sind." *Norddeutscher Rundfunk.* 17 May 2017. https://www.ndr.de/nachrichten/niedersachsen/lueneburg_heide_unter elbe/Wo-tote-Woelfe-Arbeitsalltag-sind,wolf3118.html [accessed 25 March 2018].

KORA, 2017. "Wolf Genetics." http://www.kora.ch/malme/MALME-species-comp endium/03_wolf/1_biology/genetics/genetics-wolf.htm [accessed 27 September 2017].

Krange, O. and K. Skogen. 2011. "When the Lads Go Hunting. The 'Hammertown Mechanism' and the Conflict over Wolves in Norway." *Ethnography* 12 (3), 466–489.

Krüger, G., A. Steinbrecher, and C. Wischermann. 2014. "Animate History. Zugänge und Konzepte einer Geschichte zwischen Menschen und Tieren." In *Tiere und Geschichte. Konturen einer Animate History*, edited by G. Krüger, A. Steinbrecher, and C. Wischermann, 9–34. Stuttgart: Franz Steiner Verlag.

Kurth, M. 2016. "Ausbruch aus dem Schlachthof. Momente der Irritation in der industriellen Tierproduktion durch tierliche Agency." In *Das Handeln der Tiere. Tierliche Agency im Fokus der Human-Animal Studies*, edited by S. Wirth, A. Laue, M. Kurth, K. Dornenzweig, L. Bossert, and K. Balgar, 179–202. Bielefeld: Transcript.

Kurth, M., K. Dornenzweig, and K. Wirth. 2016. "Handeln nichtmenschliche Tiere? Eine Einführung in die Forschung zu tierlicher Agency." In *Das Handeln der Tiere. Tierliche Agency im Fokus der Human-Animal Studies*, edited by S. Wirth, A. Laue, M. Kurth, K. Dornenzweig, L. Bossert, and K. Balgar, 7–42. Bielefeld: Transcript.

Latour, B. 2010. *Eine neue Soziologie für eine neue Gesellschaft. Einführung in die Akteur-Netzwerk-Theorie.* Frankfurt am Main: Suhrkamp.

Lescureux, N. 2006. "Towards the Necessity of a New Interactive Approach Integrating Ethnology, Ecology and Ethology in the Study of the Relationship between Kyrgyz Stockbreeders and Wolves." *Social Science Information* 45 (3), 463–478.

Marotzki, W. 2004. "Qualitative Biographical Research." In *A Companion to Qualitative Research.* Translated by B. Jenner, edited by U. Flick, E. von Kardorff, and I. Steinke, 101–107. London: SAGE.

Marvin, G. 2015. *Wolf.* London: Reaktion Books.

NABU: Naturschutzbund Deutschland, 2014. "Größte Gefahr: Der Mensch. 50 Wölfe auf Straßen und Schienen getötet." 28 May 2014. https://www.nabu.de/news/2014/ 05/16848.html [accessed 10 September 2017].

NABU: Naturschutzbund Deutschland, 2018. "Verkehrsopfer Wolf." 20 February 2018. https://niedersachsen.nabu.de/tiere-und-pflanzen/saeugetiere/wolf/23957.html [accessed 22 February 2018].

NDR: Norddeutscher Rundfunk, 2017. "Wölfe fressen Lämmer, keine Zäune." 18 May 2017. http://www.ndr.de/nachrichten/lueneburg_heide_unterelbe/Schaefer-Ja hnkes-Zaun-nicht-vom-Wolf-zerbissen,wolf3144.html [accessed 30 June 2017].

NDR: Norddeutscher Rundfunk, 2018a. "Lies fordert Sender für Cuxhavener Wolfs-rudel." 9 January 2018. https://www.ndr.de/nachrichten/niedersachsen/oldenburg_

ostfriesland/Lies-fordert-Sender-fuer-Cuxhavener-Wolfsrudel,wolf3338.html [accessed 22 February 2018].

NDR: Norddeutscher Rundfunk, 2018b. "Walsroder Wolf: Lies bereitet Besenderung vor." 19 January 2018. https://www.ndr.de/nachrichten/niedersachsen/lueneburg_heide_unterelbe/Lies-will-Walsroder-Wolf-besendern-lassen,wolf3356.html [accessed 22 February 2018].

Nds. MU: Niedersächsisches Ministerium für Umwelt, Energie, Bauen und Klimaschutz, 2017a. "EU stimmt höheren Ausgleichszahlungen beim Wolfsmanagement zu - Umweltminister Stefan Wenzel: Unterstützung der Nutztierhalter und Stärkung der Akzeptanz." 18 May 2017. http://www.agrar-presseportal.de/Nachrichten/EU-stimmt-hoeheren-Ausgleichszahlungen-beim-Wolfsmanagement-zu-Umweltminister-Stefan-Wenzel-Unterstuetzung-der-Nutztierhalter-und-Staerkung-der-Akzeptanz_article2407 4.html [accessed 22 February 2018].

Nds. MU: Niedersächsisches Ministerium für Umwelt, Energie, Bauen und Klimaschutz, 2017b. "Rede des Niedersächsischen Ministers für Umwelt, Energie und Klimaschutz Stefan Wenzel." 20 September 2017. https://www.umwelt.niedersachsen.de/aktuelles/p ressemitteilungen/rede-landtag-wolf-157856.html [accessed 27 September 2017].

Nds. MU: Niedersächsisches Ministerium für Umwelt, Energie, Bauen und Klimaschutz, 2017c. "FAQ." 15 November 2017. http://www.umwelt.niedersachsen.de/startseite/a ktuelles/informationen_zum_wolf_niedersachsen/Wolf-faq-134557.html [accessed 11 July 2017].

Nds. MU: Niedersächsisches Ministerium für Umwelt, Energie, Bauen und Klimaschutz, 2017d. "Informationen für Nutztierhalterinnen und Nutztierhalter." 6 December 2017. https://www.umwelt.niedersachsen.de/themen/natur_landschaft/foerdermoeglichkeiten/richtlinie_wolf/richtlinie-wolf-129504.html [accessed 22 February 2018].

NLWKN: Niedersächsischer Landesbetrieb für Wasserwirtschaft, Küsten- und Naturschutz, 2016. "Der Wolf ist zurück in Niedersachsen." http://www.nlwkn.niedersachsen.de/startseite/naturschutz/tier_und_pflanzenartenschutz/wolfsbuero/der-wolf-ist-zurueck-in-niedersachsen-144225.html [accessed 10 September 2017].

NLWKN: Niedersächsischer Landesbetrieb für Wasserwirtschaft, Küsten- und Naturschutz, (2019). "Tote Wölfe in Niedersachsen." http://www.nlwkn.niedersachsen.de/startseite/naturschutz/tier_und_pflanzenartenschutz/wolfsbuero/totfunde/tote-woelfe-in-niedersachsen-142406.html [accessed 22 February 2019].

NLWKN: Niedersächsischer Landesbetrieb für Wasserwirtschaft, Küsten- und Naturschutz, "Welcome to NLWKN." https://www.nlwkn.niedersachsen.de/service/nlwkn_international/information_english/welcome-to-nlwkn-45575.html [accessed 1 November 2017].

NLWKN: Niedersächsischer Landesbetrieb für Wasserwirtschaft, Küsten- und Naturschutz, "Das Wolfsbüro des NLWKN." http://www.nlwkn.niedersachsen.de/startseite/naturschutz/tier_und_pflanzenartenschutz/wolfsbuero/das-wolfsbuero-des-nlwkn-134954.html [accessed 27 September 2017].

NLWKN: Niedersächsischer Landesbetrieb für Wasserwirtschaft, Küsten- und Naturschutz, 2017. "Richtlinie über die Gewährung von Billigkeitsleistungen und Zuwendungen zur Minderung oder Vermeidung von durch den Wolf verursachten wirtschaftlichen Belastungen in Niedersachsen (Richtlinie Wolf). RdErl. d. MU v. 15.5.2017–26–04011/01/101- VORIS 28100, Lesefassung, Stand 06.12.2017." https://www.nlwkn.niedersachsen.de/download/109668/Richtlinie_Wolf_RdErl._d._MU_v._15._5._2017_-_Lesefassung.pdf [accessed 21 February 2018].

Ohrem, D. 2016. "(In)VulnerAbilities. Postanthropozentrische Perspektiven auf Verwundbarkeit, Handlungsmacht und die Ontologie des Körpers." In *Das Handeln der Tiere. Tierliche Agency im Fokus der Human-Animal Studies*, edited by S. Wirth, A. Laue, M. Kurth, K. Dornenzweig, L. Bossert, and K. Balgar, 67–91. Bielefeld: Transcript.

PAZ: Peiner Allgemeine Zeitung, 2017. "Naturschützer locken Touristen mit Wolfsbeobachtung." http://www.paz-online.de/Nachrichten/Der-Norden/Uebersicht/Naturschuetzer-locken-Touristen-zur-Wolfsbeobachtung-nach-Niedersachsen [accessed 10 September 2017].

Randt, J. 2017. "Touren auf Wolfsspuren." *Weser Kurier*. 21 June 2017. http://www.weser-kurier.de/region/niedersachsen_artikel,-touren-auf-wolfsspuren-_arid,1616157.html [accessed 10 September 2017].

Reding, R. 2018. "Wölfe in Niedersachsen, Bericht der Landesjägerschaft Niedersachsen e.V. zum Wolfsmonitoring." 28 January 2018. https://www.wolfsmonitoring.com/fileadmin/dateien/wolfsmonitoring.com/pdfs/Wolfsmonitoringbericht_LJN_2016_2017.pdf [accessed 22 February 2018].

Reiseland Niedersachsen, 2017. "Die Wolfstour (Süd-Route)." http://www.reiseland-niedersachsen.de/die-wolfstour-sued-route [accessed 10 September 2017].

Roscher, M. 2016. "Zwischen Wirkungsmacht und Handlungsmacht. Sozialgeschichtliche Perspektiven auf tierliche Agency." In *Das Handeln der Tiere. Tierliche Agency im Fokus der Human-Animal Studies*, edited by S. Wirth, A. Laue, M. Kurth, K. Dornenzweig, L. Bossert, and K. Balgar, 43–66. Bielefeld: Transcript.

Rosenthal, G. and W. Fischer-Rosenthal. 2004. "The Analysis of Narrative-biographical Interviews." In *A Companion to Qualitative Research*. Translated by B. Jenner, edited by U. Flick, E. von Kardorff, and I. Steinke, 259–265. London: SAGE.

Schwertl, M. 2013. "Vom Netzwerk zum Text. Die Situation als Zugang zu globalen Regimen." In *Europäisch-ethnologisches Forschen. Neue Methoden und Konzepte*, edited by S. Hess, J. Moser, and M. Schwertl, 107–126. Berlin: Reimer.

Senckenberg Forschungsinstitut und Naturmuseum Frankfurt, 2017. "Naturschutzgenetik." http://www.senckenberg.de/root/index.php?page_id=15352&preview=true [accessed 27 September 2017].

Skogen, K. 2001. "Who's Afraid of the Big, Bad Wolf? Young People's Responses to the Conflicts over Large Carnivores in Eastern Norway." *Rural Sociology* 66 (2), 203–226.

Skogen, K. and O. Krange. 2003. "A Wolf at the Gate: The Anti-Carnivore Alliance and the Symbolic Construction of Community." *Sociologia Ruralis* 43 (3), 309–325.

Skogen, K., I. Mauz, and O. Krange. 2008. "Cry Wolf!: Narratives of Wolf Recovery in France and Norway." *Rural Sociology* 73 (1), 105–133.

Tønnessen, M. 2010. "Is a Wolf Wild as Long as It Does Not *Know* that It Is Being Thoroughly Managed?" *Humanimalia* 2 (1), 1–8.

Ullrich, J. and A. Ulrich, eds. 2014. *Tiere und Tod*. Berlin: Neofelis.

Wischermann, C. 2014. "Tiere und Gesellschaft. Menschen und Tiere in sozialen Nahbeziehungen." In *Tiere und Geschichte. Konturen einer Animate History*, edited by G. Krüger, A. Steinbrecher, and C. Wischermann, 105–126. Stuttgart: Franz Steiner Verlag.

Wisentgehege Springe, 2018. "Home." http://www.wisentgehege-springe.de/ [accessed 21 February 2018].

WNON: Weidertierhalter Deutschland, 2017. "Wölfe fressen kein Gras." http://www.wnon.de/ [accessed 10 September 2017].

Wolfcenter, 2018. "Übernachtungen/Baumhaushotel." http://www.wolfcenter.de/Uebernachtungen-Baumhaushotel-160.html [accessed 22 February 2018].

Wolfsmonitoring.com, 2018. "Wolfsmonitoring in Niedersachsen." https://www.wolfsmonitoring.com/monitoring/wolfsmonitoring/ [accessed 22 February 2018].

Zips, M. 2016. "Trauriges Ende eines Problemwolfs." *Süddeutsche Zeitung*. 28 April 2016. http://www.sueddeutsche.de/panorama/niedersachsen-warum-problemwolf-kurti-sterben-musste-1.2971589 [accessed 11 July 2017].

Research material

Field notes and interview transcript, Wolfstag Springe, 20 February 2016.
Field notes, public discussion Hildesheim, 5 June 2016.
Field notes, bonfire, 12 May 2017.
Field notes, exhibition Landesmuseum Hannover, 21 May 2017.
Photos, exhibition Landesmuseum Hannover, 21 May 2017: 20170521_134003, 20170521_143307, 20170521_143424, 20170521_143517, 20170521_143535, 20170521_143613.

10 Hunting wild animals in Germany

Conflicts between wildlife management and 'traditional' practices of *Hege*

Thorsten Gieser

Introduction

Wild animals are returning to Germany: the wolf, lynx, wildcat, beaver, white stork and even the elk. Some of them are returning by themselves (wolf, white stork, elk), others are reintroduced (lynx, beaver). Although populations are still small (but generally rising), the impact of these 'new' species is already evident in various negotiations, conflicts and managerial issues regarding changing compositions of the fauna, land use and our relationship to wild animals and wild places. This return of wild animals is accompanied by a corresponding 'rewilding' discourse. In recent years, 'wilderness' has regained value as a buzzword for some conservationists – right next to other buzzwords such as 'sustainability'. From the reintroduction of species to landscape restoration or land abandonment, rewilding proponents argue – somewhat paradoxically – for managerial human interventions to create an 'untouched' wilderness, full of wild animals (Jørgensen 2015).

When following these debates, one could have the impression that managing wild animals is a relatively new phenomenon and challenge, and that its primary aim is to ensure the survival of formerly extinct species in the country. Yet, at the same time, wildlife populations in Germany seem to be at a historical peak, especially the bigger mammals, such as red deer, roe deer and wild boar. What is more, certain 'invasive' species, such as raccoons or nutria, are increasing and spreading. In addition, one should not forget that the intentional introduction of certain game species (e.g. fallow deer or mouflon sheep and, most recently, muntjac deer) also has a long history in this country. Thus, my argument is that the return of certain high-profile predatory animals to Germany (wolf and lynx in particular) has to be seen against the background of a long and complex relationship of humans and wildlife. We may follow José Ortega y Gasset's (1995) argument and propose that the principal relationship between humans and wild animals (although by no means the only relationship) has always been and perhaps still is hunting. With this hypothesis in mind, the aim of this chapter is to address how the challenge of managing wild animals is inextricably intertwined with hunting practices.

Almost 385,000 non-professional hunters in Germany are supposed to 'manage' wildlife populations, killing an average of 4.5 million game animals per year in a process officially governed by forestry and conservation agencies.[1] Hunters' representative bodies often join ecological argumentations in public discourses on the function of hunting for managing wildlife and ecosystems to justify their practice to the (mostly) critical mainstream public. On the ground, however, hunters tend to vehemently refuse being labelled as 'managers' and, instead, claim to be engaged in *Hege*, a particular 'traditional' form of stewardship that defines the hunters and their relationship with wild animals.

This relationship has been under attack for some years now. As I shall show, there is a core conflict over managerial practices which is fought out between wildlife managers, on the one hand, and 'traditional'[2] hunters, on the other. Nature conservationists, foresters and ecologically minded hunters criticise *Hege* practices as outdated, harmful and ecologically dysfunctional. The forester and hunter Wilhelm Bode (Bode and Emmert 1998) – one of the most prominent critics – argues that *Hege* practices have been geared towards the production of trophies, overpopulation of game species and, consequently, have resulted in severe ecological disbalances. In opposition to traditional hunting practices, Bode and like-minded hunters have been promoting a new 'ecological' way of hunting based on scientific knowledge (especially of ecology and wildlife biology), following rational principles and protecting animal welfare. Nowadays, this new way of hunting is promoted by the *Ökologischer Jagdverband*, founded in 1988 and with a current membership of around 2,800 hunters.[3] The *Ökologischer Jagdverband* opposes the 'traditionalist' *Deutscher Jagdverband* – founded as the *Allgemeiner Deutscher Jagdschutz-Verein* in 1875 – with about 247,000 members today. The disparity of membership numbers to the contrary, it is the new ecological hunt as a form of wildlife management that has become the norm and sets standards for many hunting practices due to its institutional basis in state institutions, such as forestry agencies and national park administrations. However, the debate on how to hunt 'properly' is still far from resolved.

My argument in this chapter is that we find differing human–animal relationships at the root of this conflict, brought about by the scientification of hunting and the (re-)introduction of the wilderness idea, now in the guise of ecology. For wildlife managers, hunting is intervention (in natural-ecological dynamics) and disturbance (of natural animal behaviour); for traditional German hunters, hunting as *Hege* is a form of relationship with animals and the land, particularly defined through care and stewardship.

We need to, firstly, look into the history of wildlife management and its relationship to hunting practices and knowledge to understand and explain the source of this conflict. On the one hand, I will show that understanding hunting as a way of 'managing' wildlife is not new in Germany but has a rather long history, dating back several centuries. What needs to be examined though, is how conceptions of management differ between ecological and traditional hunters. On the other hand, understanding hunting as a way of managing 'wildlife' is relatively new. We need to acknowledge the influence of a specific North American idea of wilderness on modern wildlife management practices and the role hunting is assigned to within this discourse. Equipped with these insights, we can move on to consider hunting in Germany and how practices of *Hege* oppose the model of wilderness and wildlife management on the grounds that hunters claim to be in a special relationship with wild animals. In conclusion, I point out potential negative implications within the agenda of wildlife management for developing environmental ethics and, instead, follow the lead of hunters to suggest that we need to think about discussing our relationship with wild animals.

Wildlife management and hunting in the past

The origins of modern, globally distributed wildlife management are often traced back to the North American conservationist, forester and hunter Aldo Leopold (see Nadasdy 2011). However, the history of wildlife management – at least for Germany – is far longer, and, most importantly, it is a history of hunting and the formation of 'hunting science' (*Jagdwissenschaft*). Wildlife surveys, measures of population control, proto-'sustainable' management of populations, hunting quotas, quantitative methods of bookkeeping, and so on, had been practised by professional hunters and hunting officials for more than 200 years before Leopold formulated his ideas in *Game Management* (1933). Hunting had undergone transformations due to an increasing bureaucratization and scientification of all areas of public life in the wake of the formation of the modern nation state in central Europe during the fifteenth to eighteenth centuries (see Pattberg 2007). Regarding hunting knowledge, this process meant a rationalization of hunting practices and the complementation of practices with a (written) scientific 'theory'. Hunting knowledge has been written down and published in books since the thirteenth century. However, these were mainly practical handbooks on the art or craft of hunting, written by experienced professional hunters to provide education for fellow hunters and apprentices. The construction of hunting knowledge as science that began in the seventeenth century had an altogether new quality.

The term *Jagdwissenschaft* was originally coined by Johann Täntzer in his treatise *Der Dianen Hohe und Niedere Jagd* (1682–1689). It referred primarily to animal biology and ethology of prominent game species and covered knowledge that most professional hunters of the time would have had. A broader conceptualized *Jagdwissenschaft* was proposed by von Fleming in his *Der vollkommene Teutsche Jäger* (1749), and he extended the science to include *Jagdgeschichte* (history of hunting) and *Jagdrecht* (hunting law). By 1839, hunting science had become an all-encompassing science and counted natural history, general history, literature, art history, technology, economics, mathematics, law, physics and chemistry among its 'assistant sciences' (see Behlen's *Lehrbuch der Jagdwissenschaften*, 1839). With the ongoing scientification of hunting knowledge, came a quantitative conception and organization of hunting practices. Various forms of accounting were introduced (see von Heppe's *Lehr-prinz*, 1751) to assist the planning of the new big hunts of absolutist hunting lords, to ensure efficiency in the use of natural resources (game), ensure the sustainability of wildlife populations and minimize damage to agriculture by game animals. This development was driven by the 'cameralist' economy of the Prussian state (and smaller German states which followed this model), a form of state-governed mercantilism that aimed at maximizing profits and efficiency in all fields of the economy, including agriculture, forestry and hunting.

With the extension of hunting knowledge to a hunting science, came the need for specialist knowledge that could not be provided by hunters themselves but had to come from academic or bureaucratic state officials. These theoretical hunting 'experts' had to be professionally educated, for example, in newly founded forestry schools (Theilemann 2004). These schools ensured the further spread of hunting science and promoted the scientification of hunting knowledge in general. This development can be considered as the beginning of a process wherein knowledge was gradually removed from local practitioners and institutionalized and standardized in the hands of theoretical 'experts'. When we look at the situation today, we find that the sciences of ecology, wildlife biology and conservation science have largely replaced the local hunters' knowledge of game animals, at least in public debates on wildlife management.

Therefore, it could be argued that wildlife management and its supporting sciences in Germany have actually developed out of hunting practices and hunting science. Up to the beginning of the twentieth century, wildlife management was one subfield of hunting. Since then, the roles have changed, and hunting has instead become part of wildlife management in the eyes of state administrators. Traditional hunters mostly resist this reversion of roles and the accompanying loss of social power regarding hunting issues. Hunting as a practice had long been central to the aristocratic state but not so any longer in a late modern society. With a loss of its former main functions (for income, means of distinction, food resource), hunting still has to find its role in today's Germany – and the hunters still struggle to find theirs (Maylein 2010).

This problematic relationship between hunting and wildlife management in Germany's history is mirrored on a global scale. The new wildlife management discourse and practice, which was initiated by Aldo Leopold (1933), builds on a new understanding of 'conservation' *with* but also *against* hunting. In his science of wildlife management, hunting was set in a framework that combined hunting with scientific theories of biology and its new subdiscipline of ecology. His inspiration in this regard came from his forestry training at the Yale Forestry School, which was dominated by German teachers (such as Dietrich Brandis and Wilhelm Schlich) and their conception of scientific forestry and hunting. Susan Flader summarized the new forestry approach:

The forester as technician was concerned chiefly with timber, but the forester in his capacity as land manager, Leopold believed, was responsible for putting the land to its highest use. That involved provision for recreational hunting as well as for the harvesting of timber and the grazing of livestock.

(Flader 1994, 66)

Leopold's management model, thus, followed in the footsteps of earlier developments in German hunting and forestry. In the preface to *Game Management*, he stated that "Game management has long been an empirical art in Europe, but the attempt to adapt that art to biological principles and to American conditions and traditions, is new" (Leopold 1933, xxxii).

As I have shown above, Leopold was wrong in his assessment of German/ European game management as a mere art or craft. His science of game management has a predecessor in German *Jagdwissenschaft*. What was new, though, was Leopold's attempt to redescribe this hunting science within a framework of the new science of ecology combined with cultural ideas of 'wilderness' into the idea of a pristine ecosystem. From Leopold's time onwards, this new science of wildlife management has become an increasingly global practice and theory which affects human–animal relationships worldwide.

Wildlife management and hunting today

Although approaches to and definitions of wildlife management are numerous, they all seem to agree on a conception of what Gisela Kangler (2009) calls ecosystem-wilderness (*Ökosystem-Wildnis*). Similar to Thomas Kirchhoff and Ludwig Trepl (2009), she argues that the science of ecology needs to be understood as a socio-cultural construction that unites scientific findings and theory with cultural values, such as ecosystem and wilderness. There seems to be an underlying correspondence between the scientific idea of 'impersonal causal forces' and the idea of a pristine nature. When combined, these ideas lead to a conception of ecosystems, for example, which are supposed to function at their best when 'natural' dynamics are left untouched by human impact – or at least, when human influence is reduced to a minimum. Paradoxically, to achieve this minimum of impact, ecosystems have to be thoroughly 'managed'. This is what William Cronon referred to as the "trouble of wilderness":

> we need an environmental ethic that will tell us as much about using nature as about not using it. The wilderness dualism tends to cast any use as ab-use, and thereby denies us a middle ground in which responsible use and non-use might attain some kind of balanced, sustainable relationship.
>
> (Cronon 1995, 85)

This can be illustrated with the example of the *Rotwildgebiet Südschwarzwald* (Red Deer Area South Black Forest). As an experiment in finding new forms of conservation areas in "cultural landscapes", the *Rotwildgebiet* offers "integrative wildlife management" that mediates between the interests of forestry, hunting, agriculture, tourism and conservation, which often conflict (for all the following information, see Suchant, Burghardt, and Gerecke 2008). Wildlife management, thus, regulates far more than wildlife; its task has broadened to encompass the negotiation of various forms of human–wildlife and human–landscape engagements. It aims at (1) regulating wildlife populations in relation to the habitat and (2) reducing 'disturbances' (i.e. human impact, especially by tourists and hunters) by (3) establishing spatial separation between human and non-human activity areas. Accordingly, the *Rotwildgebiet* is structured into a core, intermediate and marginal area, with decreasing regulations (from inner to outer) regarding human–wildlife interactions (including hunting).

We can see here how wildlife management construes hunting as intervention rather than a relationship. The key word is 'disturbance' (*Störung*). Within the core area, there are designated 'sanctuaries' (*Wildruhebereiche*), where animals are supposed to be able to live in peace, protected from human activity. The hunting season is closed all year apart from three weeks and hunting practices are generally restricted (as in the other areas of the *Rotwildgebiet*). Individual hunting is particularly restricted in favour of large-scale pressure hunts (*Drückjagden*), organized at intervals (*Intervalljagd*). Rather than allowing some individual hunters to 'disturb' wildlife throughout the year according to their own preferences, the disturbances are reduced to a minimum but then with the full force of large groups of hunters going into the area systematically in a coordinated effort for maximal efficiency.

Interestingly, disturbance through hunting refers not only to wildlife, but also to wildlife tourism. Managers have an interest in reducing hunting disturbance as it affects the behaviour of animals and, thus, the chance to observe and experience wildlife by the general public (one of the established aims of the conservation area). This interdependency between hunting and wildlife tourism influences not only when and where hunting is allowed but also transforms hunting methods. Managers in the *Nationalpark Schwarzwald* nearby experiment with a new hunting technique, the so-called *Synchron-Dublette* (Fuhr 2016a). Two hunters shoot simultaneously at two animals, (ideally) killing both at the same time, leaving no animals to witness the death and, thus, warn other animals. Thereby, animals continue visiting this site and continue to be 'experience-able' to tourists in the near future.

When considering these examples, we may conclude that hunting, for wildlife management, is a form of human impact best to be avoided or at least reduced to its functional essence: killing – and thereby regulating wildlife populations. An American example by sociologist Jan Dizard (1999) further accentuates the attitude of wildlife managers towards hunting. In a conflict over managing the local deer population involving a Metropolitan District Commission in Massachusetts and the Division of Wildlife and Fisheries, the former party planned to hire sharpshooters instead of hunters to do this job. The environmentalists in the Metropolitan District Commission "wanted *managers*, not sportsmen, loose in the reservation, people who would approach the hunt with the idea of getting a job done, not with the idea of satisfying primal urges or embodying some abstract notion of a sporting ethos" (Dizard 1999, 122). The effective and efficient administration of death as the execution of management planning is, therefore, the ideal of this type of wildlife management.

As I will show next, this conception goes against the core of the hunters' ethos – not just in the USA but especially in Germany.

Hunting in Germany

Hunting had been a common right for every 'free man' for the ancient Germans. However, from the early Middle Ages up to the German Revolution of 1848, hunting had been increasingly turned into the legal privilege of the aristocracy and the (also mostly aristocratic) professional hunters employed in their service (Maylein 2010; Rösner 2004). For centuries, the general population's only means to hunt was poaching (Girtler 1998) or, for a selected few, by being granted the hunting rights for certain 'lesser' game from a local landlord. It is only in the aftermath of the German Revolution that hunting rights were made, principally if not actually, available to all citizens. Since then, hunting has developed into a pastime mainly of the middle classes (Hiller 2003), although the influence of the aristocracy on hunting and in hunting institutions was substantial until the beginning of the twentieth century (Theilemann 2004). It could even be argued that what is commonly discussed as 'traditional' in contemporary hunting actually refers to the aristocratic habitus of hunting (which is still being practised by the remaining hunting aristocracy today).

What is significant for the hunting situation in contemporary Germany is that it is mainly a practice of non-professionals. There are about 1000 professional hunters in Germany[4] and only a few thousand hunting foresters, for whom hunting and the organization of hunts is part of their job. All other hunters hunt in their spare time, at their own cost. They have decided to become hunters despite the difficulties involved: one cannot have a criminal record, one has to undergo expensive training (min. 6 months in an informal apprenticeship with an experienced hunter connected to the *Deutscher Jagdverband* or three weeks in a recognized hunting school), with several official exams that need to be passed. Then, one either hunts as a (often paying) guest in someone else's hunting district (*Revier*) or hunts on one's own *Revier*, which may be owned or leased from landowners, often for several thousands of Euros (depending on the 'quality' of the *Revier*, i.e. its size, landscape and, most importantly, the composition of its game population). Hunters are then responsible for the game population in their *Revier*, as well as for any damage game might cause to surrounding crops in the fields.

In a way, today's hunter continues the role of professional hunters in earlier days – with the exception that they are not employed by landowners but take on their responsibility voluntarily. In other words, Germany's game population is managed to a large extent by volunteers. They have paid for their privilege to have responsibility and they see themselves as continuing the (aristocratic) hunting tradition – both factors together make them difficult partners when trying to negotiate *how* they should manage wildlife.

Hunters and animals

When discussing the hunters' relationship with animals, it is helpful to start with the German notion of a 'game animal', i.e. *Wild*. In contrast to the English term which situates animals as game in a context of the cultural practice of sports (including notions of fairness and sportsmanship, see Dizard 1999), the German term *Wild* means wild – as in wild animal. However, *Wild* does not refer to any wild animal, but only to huntable wild animals. Although they are animals defined by their freedom, they are also defined by their relationship with humans. Hence, their 'wildness' does not include being untouched by humans – as the notion of a wild animal in the common wilderness discourse might suggest.

Nowadays, German law (the *Bundesjagdgesetz*) distinguishes between *wildlebende Tiere* or *Wildtiere* (wild animals) in general and about 48 species of *Wild* (game animals) (see BJagdG 2019, §2). Some of these species (such as the lynx) are also part of nature protection laws and have a close season all year round but still remain under the management responsibility of hunters. Notwithstanding this peculiar distinction between *Wild* and wild animals in contemporary Germany's legal language, hunting knowledge 'traditionally' separates wild from domestic animals, as can be seen in a manual for hunters from the eighteenth century: von Fleming's *Der vollkommene Teutsche Jäger* (1749). Here, von Fleming elaborates the term *Wild* by further distinguishing *bestiae* from *animales* and *raptores* (using Latin terms as was usual in 'scientific' works). The *bestiae* included large, dangerous animals reserved for hunting by a king, from lions and tigers (both non-native in Germany) to bears and aurochs (both extinct now in Germany). Less dangerous, but dangerous to both humans and *Wild* nevertheless, were the *raptores*, i.e. predatory animals (*Raubwild*) such as wolf, lynx, fox and badger. The most important animals for hunters, however, were the *animales* proper, i.e. the *Wild* – subdivided into lower (*Niederwild*), sometimes middle (*Mittelwild*), and high (*Hochwild*), according to the level of aristocracy who was permitted to hunt them.

Contemporary hunting practices still recognize these categorizations of different kinds of *Wild*. Therefore, the hunters' relationship to animals is a multiple and complex one. We particularly need to pay attention to the hunters' negative relationship towards predators and their positive relationship to *Hochwild* (e.g. mainly deer species, wild boar, ibex, mouflon) and *Niederwild* (e.g. roe deer, hare, ducks, geese). Predatory animals can be either wild or domestic, but both are considered primary enemies and main competitors of the *Hoch-* and *Niederwild* and, therefore, have to be controlled through hunting, i.e. killing. On the one hand, there are the wild predators. Chief among them is the wolf – a *Raubwild* formerly hunted to extinction, as they were considered the main threat to humans and *Wild*, and which has recently returned to Germany and is spreading rapidly across the country. It comes as no surprise that hunters are among the main opponents to their return (or intentional reintroduction by animal activists, as some hunters suspect). If the major hunting magazines are to be believed, most hunters are in favour of reclassifying the wolf legally as *Wild*, so that they can be hunted again and, thereby, humans, *Wild* and livestock be protected. Popular hunting magazines, such as *Jäger*, feature wolf news and articles in every issue, presenting 'incidents' with domestic livestock, pet dogs and alleged 'attacks' on humans in Germany and elsewhere as evidence for the need to manage them through hunting and not just 'wolf-friendly' conservationists who are considered oblivious to the danger wolves pose. Most hunters I work with consider it only a matter of time before legislators 'see sense' and re-establish a proper hunting relationship concerning wolves. Until then, there are regular reports of wolves and lynx found shot dead – by poachers or, more likely, hunters turned poachers (as was the case of the 'Westerwald Wolf', which was shot by a hunter who apparently 'confused' it with a dog; see Wörner 2013).[5] As one hunter from eastern Germany told me, when dealing with wolves, you should heed the (the internationally well known) advice of the 'Three S's': *schießen, schaufeln, schweigen* – meaning shoot them, bury them and keep silent.

Related to *Raubwild* (if not worse) are stray cats and dogs that venture into the woods and become threats to the *Wild*. It is still legal today for hunters to kill stray cats that are further than 200 metres away from a settlement and dogs that are found in pursuit of *Wild* – an issue which is highly contested in German society, as cats and dogs have become cherished companion animals for many people. In the language of hunters, stray cats and dogs are *Raubzeug* (predatory 'stuff'). Thus, they are a subclass of domesticated animals and not of *Wild* (i.e. not *Raubwild*). To name them *Zeug* (stuff), shows the hunters' negative attitude towards these animals. Moreover, these animals are the only ones a hunter kills (*töten*); all *Wild* is not killed, but properly *erlegt* ('laid down'). Most hunters I have met express no mercy, no emotion and no regret in killing *Raubzeug*, although they have become sensitized to the problematics involved in killing dogs or cats.

Against this background emerges the hunters' rather different relationship to *Wild* proper, which is marked by a strong sense of care and stewardship, the so-called *Hege*.

The hunters' *Hege*

> Jeder, der mit der Waffe ins Revier geht, übernimmt also gegenüber Natur, Wild und Menschheit eine große Verantwortung. Für jeden Waidmann gilt daher: *Erst Heger – dann Jäger!* [Everybody who enters his *Revier* with a gun takes over responsibility from nature, *Wild* and humankind. For every hunter applies: first *Heger* – then hunter!]
>
> (Blase [1936] 1975, 363)

Hunting in Germany has been intimately connected with the practice of *Hege* since the early middle ages.[6] As the dictum of Count Sylva-Tarouca in his famous handbook of hunting *Kein Heger, kein Jäger!* proclaims: not a *Heger*, not a hunter! (Sylva-Tarouca 1899). The idea of *Hege* – as a hunting practice of care and stewardship – derives from the Christian, pastoralist attitude behind God's imperative for humans 'to rule over' all living beings (Genesis V.1: 26 and 1: 28) which are below them in the *scala naturae* (great chain of beings). For centuries, this imperative had been embodied by the king and his representatives and, thereby, found its way into aristocratic hunting practices. Thus, it involves a curious mix of dominion and care that had defined the aristocratic habitus – both in their relationship to their human and non-human subjects (Malinowski 2003; Theilemann 2004). So, even if wild animals were defined as wild due to their inherent freedom, i.e. not being in anyone's *possession* or under anyone's *dominion*, they were still considered to be ultimately under 'divine dominion', exercised on earth by the king. The aristocratic hunter, then, represents God as a kind of 'master of wild animals' and, thereby, exercises his right to rule over all beings. Hunting can, therefore, be described as a mediated relationship to the creator and his creation. As Oskar von Riesenthal put it in his famous poem *Waidmannsheil* (1880), in order to hunt 'properly' (*waidmännisch*), the hunter has to venerate and respect the *Wild* as a creation of the creator (*"den Schöpfer im Geschöpfe ehrt!"*).

Furthermore, *Hege* was connected with the fight against or colonialization of areas of wilderness in central Europe. In this scheme, wilderness (outside the Garden of Eden) is conceived as a part of God's creation, while, at the same time, it had to be reclaimed and turned to human ends through toil and struggle (Cronon 1995). As a place of divine punishment, ruled over by Satan, wilderness was the place for dangerous animals – the *bestiae* and *raptores*. Bears and wolves particularly were the key enemies and were ruthlessly hunted down or trapped and killed to extinction. On the level of landscape, wilderness had been gradually integrated into the domain of human influence through practices such as 'inforestation'. Whereas wild animals in the so-called wilderness were free for everyone to hunt up to the early Middle Ages, animals in 'forests' were legally claimed by the crown. And as most areas of wilderness were turned into forests by the twelfth to thirteenth century, most wild animals belonged to the crown from then on. By that time, the crown had also usurped hunting rights over communally owned land near settlements, hence hunting became illegal for anyone but the aristocracy and their professional hunters. Consequently, wild animals not subject to *Hege* practices effectively (if mainly on paper) ceased to exist (although it can be assumed that there were still large areas of forests so remote from aristocratic influence that wild animals rarely had to deal with hunters).

Etymologically, the word *Hege* derives from *Hag*: bushes and shrubs enclosing or fencing off an area of land or the fence itself (made from the shrubs of the *Hag*). Building a *Hag* was a common agricultural practice – from the Middle Ages onwards – to protect gardens and fields from cattle and other domesticated animals or could be used to keep these animals inside and, thus, to protect them from wild predatory animals, such as bear, wolf or lynx. However, fields also needed to be protected from other wild animals, mainly red deer and wild boar, which could substantially damage crops. Out in the forest, there were *Gehege*, fenced off areas of mainly woodland built to keep – otherwise free roaming – *Wild* in:

> Ein Gehege aber ist gleichsam der Ertract der Wildbahn, da nicht allein das Wildprät zu jagen mit Fleiß verschonet, und Menschen, Hunden und Raubthieren Ruhe und Friede hat, und seine Verhältnisse und Nahrung in Wäldern und Feldern überall ungehindert nehmen kann [...].
>
> [A *Gehege* is an area of game habitat where game is preserved from hunting by humans, dogs and predatory animals and, hence, enjoys peace and quiet and can feed in woods and fields without restrictions.]
>
> (Schweser 1774, 359)

Practices of *Hege*, therefore, established ambiguous domains 'betwixt and between' human society and wilderness. Similarly, the hunter was an ambiguous figure who moved in and out of this domain: hunting 'properly' (*waidmännisch*), he was protected by God when he went alone, by night, into the dark forest, far away from human settlements, fighting against 'wild beasts' (see Täntzer 1689). This representation of hunters can still be seen in the representation of poachers, as Girtler (1998) recorded it for the first half of the twentieth century.

Nowadays, religious sentiments are almost completely absent from hunting. But hunters are still hunting alone at dusk, in the night or at dawn, although in forests considerably lighter than in previous eras, never too far away from human habitation; and although the 'wild beasts' are the same as before, they are normally hunted in less dangerous and adventurous ways with a gun, from raised platforms. Nevertheless, hunters are still moving in and out of a 'grey zone' of socio-cultural control, as the German forest is not regularly occupied by the non-hunting population (except for urban forests with their recreational facilities). Given that there is no area left in Germany that one could label as 'wilderness', hunters today consider themselves as being *naturverbunden* (intimately related to nature), in contrast to the – in their eyes – *naturentfremdeten* (alienated) urban dwellers.

Yet, *Hege* consists of far more than the enclosing and establishing of ambiguous domains of hunting influence. It consists of a variety of practices of care and stewardship that Robert Blase, author of a classic handbook for modern hunters, defined broadly as follows:

> Was versteht man unter Wildhege? Alle Maßnahmen, die der waidgerechte Jäger zur Pflege, zum Schutze und zur Hebung des Wildbestandes anwendet, damit ein angemessener, artenreicher, kräftiger und gesunder

(d.h. dem Biotop angepasster) Wildbestand entsteht und erhalten bleibt. Die Wildhege darf nur in einem solchem Umfange durchgeführt werden, daß Wildschaden in der Land- und Forstwirtschaft und in der Fischerei vermieden werden. *Die Biotop-Hege ist zum wichtigsten Teil der Jagd geworden!*

[What is *Wildhege*? All measures taken by the fair hunter to nurture, protect and increase the game population in order to ensure and maintain an adequate level of a biodiverse, strong and healthy population (i.e. adapted to the biotope). The *Wildhege* must only be practiced in a way that prevents damage caused by game in agriculture, forestry and fishery. *The Hege of biotopes has become the most important part of hunting practices!*]

(Blase [1936] 1975, 373)

Translated into concrete practices, *Hege* might refer, for example, to the so-called *Hege mit der Büchse* (population control through hunting, 'with the gun'), *Hege mit dem Futterbeutel* (feeding in times of need), close seasons, protection from poaching, protection from predatory animals (*Raubwild/Raubzeug*), (re)introduction of game species and habitat improvements. Compared with these *Hege* practices, the actual hunting and killing of *Wild* take up far less time in a hunter's life. The same can be said for today's non-professional hunters – at least those with their own *Revier*. The hunters' relationship with *Wild* can, thus, be generally described in terms of a felt responsibility, of care, protection and stewardship.

The return of wild predators, such as the wolf (and lynx to a lesser degree), generally challenges this relationship and hunting practices today. For the first time in over 100 years, the *Wild* has to be protected from large predators (not just human poachers). Hunting dogs are also in danger of encountering wolves, especially during the big pressure hunts of the autumn and winter. Hunters, like their dogs, have to prepare for wolf encounters whenever they enter the forest at dusk, in the night or at dawn. Then, of course, these wild predators are new competitors for game, and hunting may, thus, become more difficult and less successful. In some wolf areas in the eastern states of Germany, for example, the (introduced) mouflon sheep have disappeared. More generally, hunters also report that the wolves' main prey species, such as red deer, roe deer or wild boar, develop new behavioural patterns and, thereby, challenge established hunting practices and knowledge. It could, therefore, be argued that *Hege* practices in the future will be increasingly concerned with protection from predators. However, given that the hunters' principal means to do so is by hunting, it would necessitate the legal reclassification of wolves as (*Raub)Wild* first.[7]

However, we should not consider the *Hege* for *Wild* as a matter only of protection. This felt responsibility does not preclude the right to rule over and hunt the otherwise wild and 'free' animals. The relationship between hunter and *Wild* is not an equal one. Ultimately, it is one of domination, not trust (see Ingold 2000). Positively enacted, the right to rule focuses on responsibility and care. Negatively enacted, it may result in *Hege* practices which serve the ends and needs of hunters at the expense of wildlife and their environments. *Hege* does not only produce ambiguous territories between 'nature' and 'culture'; it is an ensemble of practices inherently ambiguous, as it is inscribed with notions of stewardship as well as domination.

Conclusion

I mentioned earlier the rather unusual idea of American conservationists to employ sharpshooters instead of hunters. In a column of one of Germany's oldest hunting magazines, *Wild und Hund* (founded in 1894), hunter Eckhard Fuhr discusses national park management regimes regarding hunting and concludes: "Aus Nationalparks ist der traditionelle Heger ausgesperrt, nicht aber unbedingt der Jäger" [National parks exclude the traditional *Heger* but not necessarily the hunter] (Fuhr 2016b, 14).

In other words, Fuhr fears that national parks do not acknowledge the dual role of hunters as hunters and *Heger* but, instead, want to separate the sheer killing involved in hunting from everything else, leaving the hunters reduced to killers. In his eyes, this is a cold-blooded understanding of hunting which is based on the scientific reasoning behind national park policies: "Mit einer rein naturwissenschaftlichen Betrachtung, die kühl nur die Biologie der Tier- und Pflanzenarten und die ökologischen Prozesse ins Auge fasst, ist zu keinem Ergebnis zu kommen" [One cannot come to terms with a purely natural scientific perspective which considers coolly only the biology of animal and plant species and ecological processes] (Fuhr 2016b, 14).

With this chapter, I wanted to show that this is the crucial difference between conservationists and hunters: it is about how humans are generally supposed to relate, i.e. not relate to wild animals, and, consequently, how hunters in particular are supposed to reduce their 'unnatural' relationship with wild animals to killing. I believe this position is both problematic and paradoxical: it upholds and strengthens the nature–culture divide by promoting non-involvement and non-engagement and, thereby, cutting wild animals off from shared human–animal histories. We are left with a 'tourist's gaze' at wild animals from the distance as our only chance at a relationship with wild animals (as in the *Rotwildgebiet Südschwarzwald*). To put it in the words of William Cronon (1995, 90), if "wilderness leaves no place for human beings ... it can offer no solution to the environmental and other problems which confront us". We should never imagine "that we can flee into a mythical wilderness to escape history and the obligation to take responsibility for our own actions" (Cronon 1995, 90).

With this conflict in mind, we might ask: *Is there a place for human–animal engagements instead of interventions in our relationships with wild animals?* In current discourses on sustainable relationships with animals and environments, managerial models based on an ecosystem-wilderness worldview seem to prevail. Although we should be critical of the pastoralist-aristocratic habitus engrained in German hunting practices, I believe that their practices of *Hege* offer an alternative route to sustainability. With Tim Ingold (2017), we could argue that sustainability might as well be conceived as a form of care arising out of a long-term familiarity and relationship; a response-ability and responsiveness which enables hunters to live with animals in one environment shared by both. As I have shown, German hunting practices correspond to this conception of sustainability to some degree. They foster long-term relationships (through the *Revier* system) built on care and responsibility (*Hege*); however, it remains to be seen how their arguments for a relationship between humans and animals stand against arguments for a 'return of the wild' which intends to turn *Wild* into wild animals and human–animal relationships into human–animal interventions.

Notes

1 For all statistics: https://www.jagdverband.de/content/daten-und-fakten
2 When I use the term 'traditional' here, I refer to a particular group of hunters' sense of a historical relationship to a vaguely defined past. German hunting traditions have complex historical trajectories, ranging from medieval origins to later 'inventions' of the nineteenth century or early twentieth century (see Lindner 1979). Arguably, the hunters' claim to continue old traditions is not a strong one. However, it cannot be denied either that today's hunting practices are built on older practices with more or less historical depth. In this chapter, I trace the history of wildlife management more closely while offering selected insightful glimpses into the historical past of the hunters' relationship with animals and their practice of *Hege*.
3 https://de.wikipedia.org/wiki/%C3%96kologischer_Jagdverband.
4 See https://www.berufsjaegerverband.de/.
5 Between 1991 and November 2019, 407 wolves have been found dead in Germany. Forty-eight of them were illegally killed. The real number of poached wolves (including those that have not been found) is probably much higher. https://www.dbb-wolf.de/tot funde/statistik-der-todesursachen.
6 While 'practices' of *Hege* date back several centuries, the 'discourse' on *Hege* in the hunting literature started no sooner than the end of the nineteenth century (Lindner 1979).
7 Wolves have been classified as game only in the state of Saxony (since 2012). As they are still also protected through nature protection laws, hunting is still pro-hibited. But due to wolves' status as game, hunters have theoretically more to say regarding wolf management and have an obligation to become involved in wolf monitoring. Practically, this is not necessarily the case.

Bibliography

Behlen, S. 1839. *Lehrbuch der Jagdwissenschaft in ihrem ganzen Umfange*. Frankfurt: J. D. Sauerländer.
BJagdG: Bundesjagdgesetz, 2019. "Bundesjagdgesezt." Bundesministerium der Justiz und für Verbraucherschutz. https://www.gesetze-im-internet.de/bjagdg/.
Blase, R. [1936] 1975. *Die Jägerprüfung in Frage und Antwort: Ein Handbuch für Jäger*. Melsungen: Verlag Neumann-Neudamm.

Bode, W. and Emmert, E. 1998. *Jagdwende. Vom Edelhobby zum ökologischen Handwerk.* München: Beck.

Cronon, W. 1995. "The Trouble with Wilderness, or Getting Back to the Wrong Nature." In *Uncommon Ground: Rethinking the Human Place in Nature*, 69–90. New York, NY: W. Norton & Co.

Dizard, J. 1999. *Going Wild: Hunting, Animal Rights, and the Contested Meaning of Nature.* Amherst, MA: University of Massachusetts Press.

Flader, S. 1994. *Thinking Like a Mountain: Aldo Leopold and the Evolution of an Ecological Attitude Toward Deer, Wolves and Forests.* Madison, WI: University of Wisconsin Press.

Fuhr, E. 2016a. "Auch die Wildnis braucht eine Ordnung." https://www.welt.de/wissenschaft/article155170722/Auch-die-Wildnis-braucht-eine-Ordnung.html.

Fuhr, E. 2016b. "Prozessschutz auch für Hirsche." *Wild und Hund* 10, 14.

Girtler, R. 1998. *Wilderer: Rebellen in den Bergen.* Wien: Bohlau.

Hiller, H. 2003. *Jäger und Jagd: Zur Entwicklung des Jagdwesens in Deutschland zwischen 1848 und 1914.* Münster: Waxmann.

Ingold, T. 2000. *Perception of the Environment: Essays in Livelihood, Dwelling and Skill.* London: Routledge.

Ingold, T. 2017. *Knowing from the Inside.* Aberdeen: University of Aberdeen.

Jørgensen, D. 2015. "Rethinking Rewilding." *Geoforum* 65, 482–488.

Kangler, G. 2009. "Von der schrecklichen Waldwildnis zum bedrohten Waldökosystem – Differenzierungen von Wildnisbegriffen in der Geschichte des Bayerischen Waldes." In *Vieldeutige Natur: Landschaft, Wildnis und Ökosystem als kulturgeschichtliche Phänomene*, edited by T. Kirchhoff and L. Trepl, 263–278. Bielefeld: Transcript.

Kirchhoff, T. and L. Trepl, eds. 2009. *Vieldeutige Natur: Landschaft, Wildnis und Ökosystem als kulturgeschichtliche Phänomene.* Bielefeld: Transcript.

Leopold, A. 1933. *Game Management.* New York, NY: Scribner's.

Lindner, K. 1979. *Weidgerecht: Herkunft, Geschichte und Inhalt.* Bonn: Habelt Verlag.

Malinowski, S. 2003. *Vom König zum Führer: Sozialer Niedergang und politische Radikalisierung im deutschen Adel zwischen Kaiserreich und NS-Staat.* Berlin: Akademie Verlag.

Maylein, K. 2010. *Die Jagd – Bedeutung und Ziele. Von den Treibjagden der Steinzeit bis ins21. Jahrhundert.* Marburg: Tectum Verlag.

Nadasdy, P. 2011. "'We Don't Harvest Animals; We Kill Them': Agricultural Metaphors and the Politics of Wildlife Management in the Yukon." In *Knowing Nature: Conversations at the Intersection of Political Ecology and Science Studies*, edited by M. Goldman, P. Nadasdy, and M. Turner, 135–151. Chicago, IL: University of Chicago Press.

Ortega y Gasset, J. 1995. *Meditations on Hunting.* Belgrade, MT: Wilderness Adventures Press.

Pattberg, P. 2007. "Conquest, Domination and Control: Europe's Mastery of Nature in Historic Perspective." *Journal of Political Ecology* 14, 1–9.

Schweser, C. 1774. *Kluger Forst- und Jagd-Beamter.* Nürnberg: Gabriel Nicolaus Raspe.

Suchant, R., F.Burghardt, andK. Gerecke. 2008. Rotwildkonzeption Südschwarzwald: Rotwild im Südschwarzwald 2008: Konzeption eines integrativen Rotwild-Managements. https://www.waldwissen.net/wald/wild/management/fva_rotwildkonzeption/fva_gesamtbroschuere_rotwild_konzeption_schwarzwald.pdf.

Rösner, W. 2004. *Die Geschichte der Jagd. Kultur, Gesellschaft und Jagdwesen im Wandel der Zeit.* Zürich, Düsseldorf: Patmos.

Sylva-Tarouca, E. 1899. *Kein Heger, kein Jäger! Ein Handbuch der Wildhege für weidgerechte Jagdherren und Jäger.* Berlin: Paul Parey.

Täntzer, J. 1682–1689. *Der Dianen Hohe und Niedere jagtgeheimnueß.* Kopenhagen.

Theilemann, W. 2004. *Adel im grünen Rock: Adliges Jägertum, Großprivatwaldbesitz und die preußische Forstbeamtenschaft 1866–1914.* Berlin: Akademie Verlag.

von Fleming, H. F. 1749. *Der vollkommene Teutsche Jäger.* Leipzig.

von Heppe, C. 1751. *Aufrichtiger Lehrprinz oder Praktische Abhandlung von dem Leithund, als dem Fundament der edlen hirschgerechten Jaegerey.* Augsburg.

von Riesenthal, O. 1880. *Waidmannsheil.* http://www.int-st-hubertus-orden.de/html/oskar_von_riesenthal.html.

Wörner, F. 2013. "Wölfe im Westerwald: Verfolgt bis in die Gegenwart – Ein Plädoyer für Akzeptanz. Niederfischbach: Ebertseifen Lebensräume e.V./Tierpark Niederfischbach e.V." https://ebertseifen.de/download/woelfe-im-westerwald-verfolgt-bis-in-die-gegenwart/.

11 Ways of speaking, responsibility and the animals 'of the forest' in Northwest Russia

Laura Siragusa

Introduction

On 1 November 2010, I visited some elderly Veps who live in Toižeg (Russian (R.) Drugaya Reka), a village neighbouring the largest settlement, Kalaig (R. Rybreka) (with its 400–450 permanent residents) in the Republic of Karelia, Russia (Figure 11.1). I had arranged to go and visit Marina Ivanovna (pseudonym) (ca. 60) who lived in a large wooden house right at the end of the village before the stretch of forest, which separates Toižeg and Kalaig, begins (Figure 11.2). I found her talking on the phone, switching erratically between Vepsian and Russian, both of which she speaks fluently. Indeed, most elderly villagers can master their heritage language, Vepsian, and Russian.

It was clear that she was in a state of shock as she was recollecting how a bear had attacked her only a couple of hours before when she was on her way to Kalaig. She had decided to pay a brief visit to some relatives and had got onto her bike in haste. However, after a couple of turns she saw that a bear with two cubs was crossing the road which links the two villages; thus, it was making its way into a less populated part of Kalaig (Figure 11.3). Taken by surprise by the quick-paced Marina Ivanovna, the bear reacted promptly by mounting towards her and "exposing its teeth quite aggressively", as she later recollected. In her irregular talk on the phone, I understood that a passing car had rescued her from the attack. "It's a miracle that I'm alive!" she kept repeating.

We exchanged only a few words that day, since she wanted to be left alone to regain control of herself. I came to see her again after a couple of days to check how she was doing. This time, she appeared more composed and was able to recollect the episode of the bear more calmly by also providing her own interpretation of the event. She spoke Russian to me, possibly because my Vepsian skills were rather limited back then and because my position as a researcher might have suggested links between my own figure and the political institutions (cf. Sayer 1984). In her opinion, not only bears, but also other animals 'of the forest' (e.g. wolves) had made their appearance in the villages more frequently in recent years due to a continuous depopulation of the rural settlements. She explained that the insufficient investment in the villages by the regional (i.e. Karelian) and federal (i.e. Moscow) administrations forced people to leave and look for their luck in larger urban centres. Consequently, the less populated villages had become an appealing destination for the animals 'of the forest'. In her interpretation of the events, she did not mention that they had experienced an exceptionally hot summer in 2010 and the subsequent fires forced the animals to go and look for food closer to those areas occupied mostly by human beings. In this sense, the presence of the bear in the stretch of forest between Toižeg and Kalaig should not have been surprising. Instead, she blamed the insufficient investment and overall disinterest in rural areas by the political bodies and the consequent depopulation of the Vepsian villages.

What surprised me from the discussion I had not only with Marina Iva-novna but also with other villagers from Kalaig and villages in the Leningrad and Vologda oblasts was not that responsibility for the behaviour displayed by the animals 'of the forest' was attributed to different actors, but that such attributions corresponded to different *ways of speaking*. In Vepsian, the phrase the animals 'of the forest' is translated as *mecživatad*, which has lat-erally come to stand for 'wild' animals – a distinction of 'wild' and 'domestic' animals which does not comply with the more traditional Vepsian categor-ization that classifies animals depending on the territory in which they usually dwell (Vinokurova 2006). *Mecživatad* are those animals which ordinarily inhabit the territory covered by the forest, swamps and lakes, and are not found within the boundaries of the *külä* (extended village in Vepsian). On the

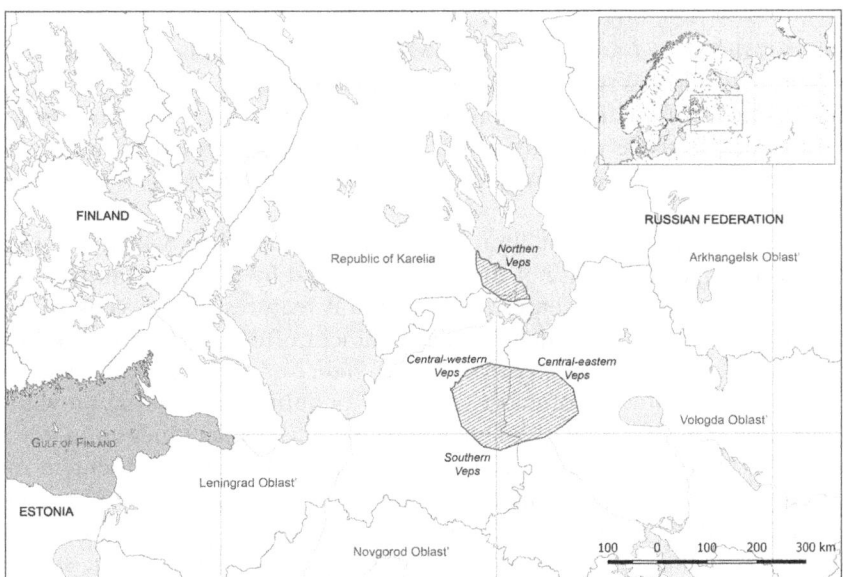

Figure 11.1 Map of the territory with Vepsian settlements (striped sections) in North-west Russia, adapted by Arch. Alessandro Pasquini

Figure 11.2 Kalaig (R. Rybreka). This is the main road, which stretches out through the forest to Toižeg (Drugaya Reka)

contrary, *kodiživatad* are the animals which dwell within the *külä* and, thus, are otherwise understood as 'domestic' animals. I have here opted for the phrase animals 'of the forest' to match the local terminology; nonetheless, this phrase indicates those animals which are otherwise regarded as 'wild' and whose movement between human settlements and the wilderness is considered problematic.

Ways of speaking are not limited to specific narratives, which the local villagers share, as is the case of the story presented by Marina Ivanovna, but also encompass different genres and structural features of the Vepsian language (cf. Duranti 2003). Such language choices often conform to the local binary distinction between *mecživatad* and *kodiživatad*. Employing one way of speaking or another is often the result of a continuous negotiation with a dynamic and always situated language ecology, i.e. a phenomenon in which language is used dynamically and in response to the forces present in one place at a specific time, including language ideologies (Garner 2004; Mühlhäusler 1996; Schieffelin, Woolard, and Kroskrity 1998).

Both *mecživatad* and *kodiživatad* can display an odd behaviour, such as making an appearance in the village, despite being usually located in the forest, and this prompts Vepsian villagers to engage either with a certain genre and structural feature of their heritage language or in a narrative

Figure 11.3 The stretch of road that links the Vepsian villages Toižeg (indicated in Russian by the signboard) and Kalaig

that is often highly politicized and expressed in the Russian language. The genre I am referring to is Vepsian *puheged* and *vajhed/pakitas* (enchantments, spells, verbal charms), and the structural feature of the language is the translative nominal case when used as an omen. By problematizing human–animal relationships through a linguistic lens, I can evidence how verbal art and bilingual practices allow people to apportion responsibility for oddities in the behaviour of the animals to different actors, be they human or non-human.

This language-based approach adds a new angle to social research conducted on 'wild' animals. The territorial misplacement of the 'wild' animals/ animals 'of the forest' reveals agency on the way people speak, how they engage with the environment and, consequently, make sense of it. Our anthology motif, the 'return of the wild', gains more depth when we understand how it is constructed linguistically. Language enables us to appreciate in more depth how people organize their world and their relationships with other-than-human beings within a diachronic and relational frame. Language not only manifests relationally and reflects what is going on in the world but also contributes to its formation and the consequent symbolism acquired (cf. Austin 1975).

A diachronic sketch of Vepsian language ecology

Veps, a Finno-Ugric minority in the Russian Federation, live in a Northwest Russian territory that they share with other nationalities, such as Russians, Karelians and Ingrians. Although traditional Vepsian settlements cover a rather compact territory, they are currently administrated by three distinct regions: the Republic of Karelia, and the Leningrad and Vologda oblasts (Puura 2012). Such administrative separation is not a recent political intervention, as the Vepsian population was split into two different administrations in the 1600s when Russia and Sweden fought over the territory around Lake Ladoga (Kolesov et al. 2007). At the time, part of Veps came under the Novgorod administration and part under Olonets. This administrative division provided the substratum upon which the Soviets later divided and managed the territory covered by Vepsian settlements and that corresponds to the current situation.

Traditionally, Veps have been living in rural areas. At least this can be said for the period which preceded the Stalinist era. Living at the periphery of the three administrative regions was soon to be challenged by the assimilation policies, which saw many Veps move from the villages to the cities, such as Petrozavodsk in Karelia, St. Petersburg (former Leningrad) in the Leningrad Oblast and Babaevo in the Vologda Oblast (Strogal'shchikova 2014). An example of this is the policy of *liquidation of the villages without prospects* launched by Khrushchev at the Twenty-Second Party Congress in 1961 (Yegorov 2006). This policy classified the villages into two categories: *with prospects* and *without prospects*. Those villages regarded as being without prospects stopped receiving any investment for public services and infrastructure (Kurs 2001, 73). It is not surprising that many Veps to whom I spoke had moved to the city where more jobs were available. This often coincided with dropping their heritage language, Vepsian, and embracing Russian to cover most domains of use.

Up to now, Vepsian villagers (especially those older than 60) have tended to be bilingual in their heritage language, Vepsian, and in Russian. However, those who had moved to the city at an earlier stage of their life found it easier to switch to Russian completely; and even though many stated that they understand Vepsian, they often admitted that they could not speak it. The Vepsian language belongs to the Finnic subgroup of the Finno-Ugric languages, together with Finnish, Estonian, Karelian, etc., and comprises three main dialects (Puura 2012; Zaitseva 2016). Veps who speak the northern dialect can be found in the Southeast part of the Republic of Karelia. Kalaig (R. Rybreka) is one of the villages situated in this northern area. Veps who use central dialects live in the Babaevo and Vyterga districts of the Vologda Oblast and in the Podporozh'e, Tikhvin and Lodeynoe Pole districts of the Leningrad Oblast (Setälä and Kala 1951). The southern-dialect speakers live in the Boksitogorsk province of the Leningrad Oblast. These dialects are considered intelligible.

Vepsian is classified as severely endangered by UNESCO.[1] In the late 1980s, Vepsian scholars and activists working at the Academy of Sciences in Petroza-vodsk prompted a revival movement with the main aim of tackling social inequality and guaranteeing investment in rural areas (Klement'yev, Kozhanov, and Strogal'shchikova 2007). Admittedly, this was a time when other revival movements sprang up all over Russia (and not only Russia, for that matter) (Vakhtin 2001), and many enthusiastic activists from different parts of the world took on the responsibility for preserving and promoting their own heritage lan-guages (cf. Evans and Sahnoun 2002; Grenoble and Whaley 2006; Henne-Ochoa and Bauman 2015). Despite their enthusiasm and enduring work, the long-established geopolitical divide began to hinder some of the Vepsian activists' efforts. While the regional administration of the Republic of Karelia began to support their work (at least partly), the legislative measures taken there did not always reach the other two administrative regions. Between 1992 and 1993, the Vepsian villages of the Republic of Karelia, Šoutjärv' (R. Shelto-zero), Kalaig (R. Rybreka) and Šokš (R. Shokhsa), obtained national status, which led to the establishment of the *Vepsian national district* in 1994 and the introduction of Vepsian language education at school (Strogal'shchikova 2004). Such initiatives were not possible in the other regions. A continuous dialogue with the regional powers in Petrozavodsk led to a political awareness among northern Veps that, I argue, is not as prominent in the two other regions where Vepsian settlements are located. Northern Veps are more likely to incorporate a political tone in their narratives, and this can also extend to the behaviour of the animals. Instead, such a narrative has only partly reached the villages in the Leningrad and Vologda oblasts, where the locals appear to engage less in poli-tical discussions and not to attribute the behaviour of the animals or other beings to the political powers' possible wrongdoings.

Given the momentum and impetus of the revival movement in the Republic of Karelia, it may appear paradoxical that central Veps tend to speak their heritage language more frequently than northern Veps. This might be because northern Vepsian villages have attracted workers from other parts of Russia (and previously the Soviet Union), thanks to the robust extractive industry there (Davidov 2013; Varfolomeeva 2016); thus, this has necessitated the use of Russian as a lingua franca (cf. Kaiser 1994). Contrarily, the locals mostly engage in Vepsian in such villages as Pondal (R. Pondala) in the Vologda Oblast, where I returned twice for extended fieldwork.

On the one hand, I do not feel comfortable marking a clear-cut distinction between northern and central (possibly southern) Vepsian *ways of speaking* concerning human–animal relationships, since they certainly vary depending on the level of the individual participation in the revival movement and the gen-eration one belongs to. On the other hand, I still believe that it is important to address such diachronic political and socio-cultural developments, since they retain agency on those practices which are shared (including language) and with which northern and central Veps engage. I often witnessed political accu-sations made by the locals in Vepsian northern villages. Here, the political institutions are often held responsible for anything that does not match the locals' expectations (Puura and Tanczos 2016). I did not witness such a widespread attitude in those central villages where I conducted research, especially Pondal (R. Pondala) in the Vologda Oblast, which is further away from the influence of larger centres and cities. This behaviour extends to human–animal relationships.

Northern Veps: responsibility in political narrative

In the autumn of 2010, I used to visit Larisa Petrovna (pseudonym) (ca. 30) who lived in Kalaig (R. Rybreka) and worked as a teacher of Vepsian at the local school. She helped me to settle into the village by offering me a place to stay at her grandmother's house and introducing me to some of the locals. Late one evening, she invited me to sleep over instead of going to her grandmother's house. She warned me that it was not advisable to be out when it was dark, as some of the other villagers had seen wolves hanging out near the village. She recounted how only a few days back, her niece went to gather some water from the well and spotted a few wolves in the distance. When I discussed the episode with some of the other teachers at school, their narrative already sounded familiar to me as the accusation turned towards the lack of investment in the village by the federal, regional and local authorities. According to many, the settlements were less populated than before and the newly-created empty space caused concern, since the animals 'of the forest' were making their way into the villages, thus appropriating some of their parts and reducing the living area covered by humans.

The *mecživatad* are supposed to dwell in the forest and not to gain territory in the area inhabited by humans and *kodiživatad*. As indicated above, the categorization of 'wild' and 'tame' does not apply to the Vepsian traditional categorization of the animals (Vinokurova 2006). Veps make a distinction between those animals they share their settlement with and those which usually reside away from it. In the different locales, the animals are supervised by different masters, which can be either human or non-human. On the one hand, the village is ruled by human beings and other, spiritual entities, which the Veps are expected to respect by maintaining a positive language and neither swearing nor screaming. A couple of these spiritual beings are the *pertinižand* (host of the house and the land where the house is built) and the *kül'betižand* (host of the Vepsian sauna). On the other hand, the forest has its own spiritual master, the *mec(a)ižand* (*mecemag*), with whom humans usually interact when entering his territory or when negotiating a deal (Arukask 2002). It appears that movement of the animals between these two territories, the human settlement and the forest, can cause disruption from the more ordinary *status quo*, which can be re-established by verbal interaction between the territorial masters and human beings.

Therefore, the presence of the animals 'of the forest' in their village was particularly upsetting for the inhabitants of Kalaig (R. Rybreka). The

political narrative shared by many of the villagers in this northern Vepsian territory reveals a direct and active engagement with the ecology in which the locals
live and the main forces which dominate there. The rather recent history of
political interaction with the political bodies in Petrozavodsk, that I presented in
the section above, allows us to attend to new forms of resistance and antagonism. Such forms of resistance are comprised in certain narratives which people
express mostly in the Russian language. Admittedly, my presence as a researcher
interested in the Vepsian language also influenced this specific way of speaking,
as I must have represented an easy way to reach out to the bodies of power and,
as many stressed, "They [i.e. administrative bodies] may listen to you more than
they listen to us!" (author's fieldnotes, 2010; Sayer 1984).

Noticeably, I did not come across such a narrative pertaining to non-
human animals in the trips that followed this one and where I visited mostly
central Vepsian villages between 2011 and 2015. This might be due to a
combination of factors: first of all, I had become more fluent in Vepsian and I
could better understand more nuances expressed in the villagers' heritage
language; secondly, my last trip focused on human–animal relationships,
which possibly prompted more traditional ways of speaking about them; and,
most importantly, the locals in those central Vepsian villages demonstrated
that they engaged less with the authorities and took more responsibility for
any occurrence which took place there.

Responsibility in *verbal charms*

There are other Vepsian *ways of speaking* which reveal how relationships
between human and non-human beings are constructed and negotiated. One
of those is the employment of *puheged* and *vajhed/pakitas* (enchantments,
spells, verbal charms in Vepsian), which are conventionally used, for example,
to cure someone who has fallen ill or been bitten by a snake, to find lost
cattle/people in the forest or swamps or to choose the land on which to build
a new house (Vinokurova 1988). The word *puheged* is etymologically connected to the verb *puhuda* (to blow) and its use attends to human-to-human
and human-to-non-human relationships and health; whereas, *vajhed/pakitas*
means 'specific words' and often expresses a humble request in order to be
granted a favour from the territorial masters, non-human beings with whom
Veps share their territory. Not everyone has the capacity to perform those
verbal charms; only those villagers called *tedai* [the one who knows the way]
or *noid* [sorcerer] in Vepsian, and some other villagers who have been
instructed in such practices (Vinokurova 1988). As the Vepsian word *puheg*
(singular of *puheged*) is etymologically connected to the Vepsian words
puhuda [to blow] and *puhutuz* [whiff of wind] (cf. Hickey 2013), this leads me
to interpret *puheged* as a way to manipulate social order through blowing.

During my fieldwork in Pondal (R. Pondala) in the Vologda Oblast and
Nemž (R. Nemzha) in the Leningrad Oblast, I managed to discuss the use of
enchantments with some of the local villagers who still employ them.

Gathering the *puheged* and *vajhed/pakitas* proved to be a rather challenging task, since many agreed that once the powerful words are shared, the *tedai* will not be able to perform them again (Kurets 2000; Lavonen 1988). Therefore, such wise individuals pass their knowledge on very cautiously, and this usually happens in their old age when they perceive a loss of strength. Choosing who should be the receiver of such knowledge also requires some consideration, as one would ideally ensure that the enchantments will be used to help others. Despite these challenges, a few villagers happily shared a few verbal charms with me and some of their stories about them. They told me how some relatives were cured of hernia or the poison from snake bites and how some old ladies in the village used to find cattle or people who were lost in the forest.

I would like now to bring to your attention to the verbal charms that Veps use to engage with the territorial masters in order to negotiate a deal regarding the animals. I will show how the misplacement of *kodiživatad* in the forest and the consequent use of those charms to bring them back home place the responsibility on the villagers and territorial masters rather than on the institutional powers. Those villagers who retain such an ability are held responsible for restoring order from a situation of undesired chaos and it is their heritage language which enables them to do so. Russian is also used in some of the enchantments, although not as frequently. Indeed, long-term contact between Russians and Veps has also allowed for an exchange of practices and systems of value (Pugh 1999). It is, however, understood that the territorial masters will play a crucial role in the fulfilment of such requests. What is also interesting is that in the case of the return of lost cattle, the movement of the animals is opposite to the one described in the section above: while in the case of Kalaig (R. Rybreka), it was the *mečživatad* ('wild' animals) which showed up in the human settlements, here, it is the *kodiživatad* ('domestic' animals) which have become lost in the forest.

The following verbal charm was used to attract the lost livestock back to the village. Valentina Kuznetsova, director of the phono-archives at the Academy of Sciences in Petrozavodsk, kindly offered it to me and Olga Zhukova transcribed it.

In 1981, the researchers Nina Onegina, Anna Kosmenko and Vera Mal'chi interviewed Aleksandra Kalinina who was 72 years old at the time (RAN 19:2662b:25). The interview took place in Pondal (R. Pondala), a central Vepsian village. Here is what Aleksandra Kalinina told the researchers:

> Mändas mecha, kumardasoiš, gostincoikš otoutas liibäd, kromaižen čapetas bokaspäi, toižen – toižespäi, siplitadas solad. Pandas ühthezo. Pandas saharad supalaižen, pandas rusttad kumačud loskutaižen, pandas tabakon sigarkaižen, pandas spičkaižid. Nu, i mändas i pakitas. Mändas mugoman puhudennoks, oliž miše kuume tesarašt, kuume tesarad om, huras kädes, ladvaižes kuume oksad duužen uuda. Nu i ladvaižehe. Tesaraiže kumardase vast peiväižele, ii kuna päiväine astub päivau, a vast päivha, ehtkoižespäi, sinna, homesbokha. Kumardasoiš i pakitas:

They go to the forest to worship and bring bread as an offering. They take the bread, cut off the crust and add a pinch of salt. They add [the salt] together. They add some sugar, lay out a red calico cloth, some tobacco from cigarettes and some matches. Then they walk and ask. They walk until they find a small tree with three intersections. There has to be three intersections, and on the left hand, there should be three branches on the top. At the intersection, they bow. They bow away from the sun, that is, not where the sun is, but away from it, towards the East. They bow and say:

Mecižand da mecemag, kazakad da lapsuded, babuško, deduško, abutagat mini pördutada živataine. Otkat neno tomaižed. Tomaižed otkat, tiile, živataine ningimale rababožjale. Mina postupimoi, povinimoi, laimoi, oigenzin, nu i kaik.

Host and hostess of the forest, workers and children, grandmother and grandfather, help me to return the livestock. Take these offerings. Take the offerings for yourselves and leave the cattle to this servant of God. I gave up, apologized, confessed, quarrelled, sent, and that is all.

Pandas [...] kädelo, kuume haškud jäl'gehepäi. Ii käroukoi mod ezile. Kuume haškud haškeitas, a potom modon ezile käroutas. Kodihe kuni asttas, ka saukoi ni-kuna ii. I toižuu päivuu živataine ob'azatel'no lüüduse.

They put [something] in the hand and take three steps backwards. They do not turn forward. Only after three steps, they turn forward. They do not talk on their way home. And the next day the cattle will definitely be found.

I do not want to state that such verbal practices only occur among central (and possibly southern) Veps, since admittedly in 2010 I did not actively look for the use of such genre among northern Veps when my research had only just started. However, I often heard northern Veps themselves claim that the Vepsian heritage language and traditions are better preserved in this more central and southern part of the Vepsian territory. This observation seems to be backed up by the fact that most enchantments which I was provided with during my archival work at the phono-archives were also gathered mostly in this area. What is of interest, however, is that during my fieldwork in those villages, such as Pondal (R. Pondala), the villagers did not make any reference to the political powers and did not blame them for the movement of the animals in or out of the village. On the contrary, by using verbal charms and the translative case, which I present in the next section, they demonstrated that they charged themselves and the territorial masters with full responsibility for the behaviour of the animals which did not appear to be entirely in control of their own lives, as they were constantly supervised either by humans or the territorial masters.

Responsibility in the translative case

In Pondal (R. Pondala), the villagers also demonstrated that they did not charge political bodies with responsibility after the animals (in this case both *kodiživatad* and *mecživatad*) displayed unusual behaviour. That materializes in the use of the translative case and other morpho-syntactic forms which cover similar functions but which I decided not to present in this chapter. In Vepsian, the translative case is formed by adding the suffix *–ks* to the root of the word and indicates not only a change of state, characteristic also of Finnish and Estonian, but also the possibility of making a prediction. Exemplarily, "Vedos pagištas čto, osobenno ezmäižen kerdan kulod kägoin g'ogen, vedon tagapei – značit *kündlikš*" [They say that if the cuckoo cries across water, especially if you hear it for the first time, it will bring tears] (author's fieldnotes, Pondal, 2013). The word *kündlikš* [it will bring tears] indicates a prediction and is provided in the translative case. By using this nominal case, Veps demonstrate that they observe the environment, any unusual event and the behaviour of the animals very closely, and this, therefore, prepares them for the possible consequences these bring about.

When the animals display unusual behaviour, such as flying towards windows, houses or even people, approaching the village from the forest or finding themselves near human settlements where their presence is not expected, Veps mark such oddity by engaging in this structural feature of the language (or those with the same function). Employing the translative case allows them to interpret such omens and prepare themselves for future openings. The animals and other non-human beings can return to the village also through dreams, for example, "Uniš näged. Kala ii *hüväks*" [If you see a fish in your dreams, it is not a good sign] (author's fieldnotes, Pondal, 2013).

The use of the translative case enables us to make a further claim regarding human–animal relationships and how responsibility is socially distributed in language. Once again, neither political nor institutional powers take part in these relationships that, on the contrary, concern the villagers and the other non-human inhabitants of this rural and forested territory intimately. In the case of the translative case, the villagers receive a message from the animals which forces them to heighten their senses and take responsibility for their response and reaction.

Language responsibility

The discussion above allows me to advance a convoluted definition for language responsibility and to branch it out in two main directions which correspond to the following questions: how is responsibility shown in ways of speaking, be they narratives, genres or structural features of the language, regarding human–non-human relations? And, what does this add to research done on 'the return of the wild'?

The Vepsian ways of speaking and allocation of responsibility appear to comply with the idea that responsibility is a social exercise, as it fosters relationships between humans, non-human beings and the broader environment (cf. Hannon 2010; Peltonen 2010). This chapter focuses on "the production of meaning [...] through dialogue and interaction which mediate between linguistic forms and their interpretations" (Hill and Irvine 1992). Indeed, Veps allocate responsibility differently in relation to the language ecology in which they live. As Dan-Cohen (1992, 959) puts it: "the individual is not free (free will) to decide something [...], but there is always a negotiation with other forces". Northern Veps tend to colour their narratives with a political shade, as they have long engaged with institutional bodies, and this allows them to accuse the institutions of negligence and carelessness; this appears to be less frequent (if at all in some places) in the villages of the Leningrad and Vologda oblasts, where language is normally used in relation to the human and non-human entities dwelling there. These entities are held responsible for one another, and Veps reinforce relationships with and within the environment through verbal art and certain structures of the language.

In fact, the appreciation of responsibility as a social act has already been enclosed in the etymology of the word *responsibility* itself, which precludes a response to a question/prompt and, consequently, some kind of interaction. The notion of responsibility retains a sense of obligation (Bernstein 1995, 14) which, thus, strengthens social binding.[23]

Responsibility in language, therefore, can transcend an individual's intentions and indicates that encounters have taken place and social binding reiterated, either harmoniously or not (cf. Du Bois 1992). This position enters into friction with speech-act theory scholars, such as Austin (1975) and especially Searle (1969), who often identify meaning with the intentions of the individual speaker in order to express certain beliefs or bring about certain changes in the world. On the contrary, in this article, the notion of responsibility in language matches the one suggested by Pyyhtinen and Tamminen (2011, 137), i.e. responsibility in language might be understood as "a product of the coming together and interchange of both human and non-human elements". I concur with Duranti (1992) and his presentation about the Samoan *fono* (meeting), where he indicates that what matters is the allocation of responsibility, and not necessarily the intentionality behind it. Only to a certain extent does responsibility in language mean acting with a rational purpose in mind, since people have to mediate their position in the ecology where they manifest language.

In the case of politicized narratives, northern Vepsian villagers hold the political institutions responsible for oddities in the behaviour of the animals; in the case of the enchantments, Veps take the responsibility themselves by negotiating with equally powerful entities (territorial masters); in the case of the translative case, Veps receive input from non-human animals, which we can interpret as an 'encounter', and respond to that prompt by engaging in a certain way of speaking and by heightening their senses. The movement of the animals in and out of human settlements expressed in ways of speaking indicates such interaction and responsiveness to the ecology in which people live. Adding such a language-oriented investigation helps us identify how people interpret such movement and oddities in the behaviour of the animals within the socio-political conditions in which they manifest and experience language. These experiences vary in time and respond dynamically to the conditions in which one lives. Ways of speaking regarding other-than-human entities cannot be extrapolated from the dominant language ecology and a long-established relationship among human and non-human entities. It is not surprising that responsibility for the behaviour of the animals is equally distributed according to the dominant systems of values and habitual practices which are shared by the local dwellers.

Conclusion

With this chapter, I wanted to bring your attention to *ways of speaking* about non-human animals and how unusual behaviour (including their movement to or from the villages) can be reflected in the narratives, in the genre employed or in the structure of the language. These language choices are made in relation to the language ecology in which the speakers find themselves. Depending on the dominant ecology, people also decide to carry more or less responsibility for the behaviour displayed by non-human animals.

The presence of the government in the northern Vepsian villages is felt more strongly, and when the animals 'of the forest' move to the inhabited settlements, Veps often acquire a political tone in their speech. The animals 'of the forest' become part of an accusation of the political powers for a lack of investment in rural areas. The institutional bodies are charged with a responsibility which encompasses the anomalous behaviour displayed by the animals. In central (and possibly southern) Vepsian villages, responsibility is usually equally distributed among people and other non-human entities (territorial masters). The allocation of responsibility also appears to match certain bilingual practices: when the discourse rotates around political issues, the villagers tend to embrace Russian rather than their heritage language, Vepsian. In a more politicized ecology, such as that in northern Vepsian villages, people tend to address social and political issues in their speech and such verbal practices extend to non-human animals. Veps tend to distribute full responsibility among themselves and the territorial masters for the movement of the animals in and out human settlements in areas which are least influenced by political rhetoric of power inequality.

Such an analysis allows us to place the motif of the present anthology, 'the return of the wild', within a language-based frame and to appreciate its relational and dynamic configuration in more depth. The 'return of the wild' discourse cannot be extrapolated by the ecology in which it occurs and bringing language to the centre of the analysis enables us to place it better both socially and diachronically.

Notes

1 See: http://www.unesco.org/languages-atlas/index.php
2 See http://www.etymonline.com/index.php?term=responsible [accessed 31 March 2017].
3 Paradoxically, social binding also results from concealment practices and gossips which implicitly disdain direct confrontation and open discussions (Abrahams 1970; Jones 2014).

Bibliography

Abrahams, R. D. 1970. "A Performance-Centred Approach to Gossip." *Royal Anthropological Institute of Great Britain and Ireland* 5 (2), 290–301.

Arukask, M. 2002. "The Spatial System of Setu Kalevala-Metric Lyrical-Epic Songs." *Tautosakos Darbai* XVII(XXIV), 46–64.

Austin, J. 1975. *How to Do Things with Words*. 2nd edition. Cambridge, MA: Harvard University Press.

Bernstein, R. 1995. "Rethinking Responsibility." *The Hastings Center Report* 25 (7), 13–20.

Dan-Cohen, M. 1992. "Responsibility and the Boundaries of the Self." *Harvard Law Review* 105 (5), 959–1003.

Davidov, V. 2013. "Ecological Tourism and Minerals in Karelia: The Veps' Experience with Extraction, Commodification, and Circulation of Natural Resources." In *The Ecotourism-Extraction Nexus: Political Economies and Rural Realities of (Un)Comfortable Bedfellows*, edited by B. Büscher and V. Davidov, 129–148. London: Routledge.

Du Bois, J. 1992. "Meaning without Intention: Lessons from Divination." In *Responsibility and Evidence in Oral Discourse*, edited by J. Hill and J. Irvine, 48–71. Cambridge: Cambridge University Press.

Duranti, A. 1992. "Intentions, Self, and Responsibility: An Essay in Samoan Ethnopragmatics." In *Responsibility and Evidence in Oral Discourse*, edited by J. Hill and J. Irvine, 24–47. Cambridge: Cambridge University Press.

Duranti, A. 2003. "Language as Culture in U.S. Anthropology Three Paradigms." *Current Anthropology* 44 (3), 323–347.

Evans, G. and M. Sahnoun. 2002. "The Responsibility to Protect." *Foreign Affairs* 81 (6), 99–110.

Garner, M. 2004. *Language: An Ecological View*. Oxford: Peter Lang.

Grenoble, L. A. and L. J. Whaley. 2006. *Saving Languages. An Introduction to Language Revitalization*. Cambridge: Cambridge University Press.

Hannon, P. 2010. "Collective Responsibility." *The Furrow* 61 (6), 331–338.

Henne-Ochoa, R. and R. Bauman. 2015. "Who Is Responsible for Saving the Language? Performing Generation in the Face of Language Shift." *Journal of Linguistic Anthropology* 25 (2), 128–150.

Hickey, R., ed. 2013. *The Handbook of Language Contact*. Malden, Oxford: Wiley-Blackwell.

Hill, J., and J. Irvine, eds. 1992. *Responsibility and Evidence in Oral Discourse*. Cambridge: Cambridge University Press.

Jones, G. 2014. "Secrecy." *Annual Review of Anthropology* 43, 53–69.

Kaiser, R. 1994. *The Geography of Nationalism in Russia and the USSR*. Princeton, NJ: Princeton University Press.

Klement'yev, E., A. Kozhanov, and Z. Strogal'shchikova, eds. 2007. *Vepsy: modeli etnicheskoy mobilizatsii. Sbornik materyalov i dokumentov.* Petrozavodsk: izdaniye osushchestvleno pri finansovoy podderzhke sekretaryata Barentseva.

Kolesov, A., E. Kotkin, M. Goldenberg, V. Gromov, A. Gromtsev, V. Landgraf, M. Skripkin, and V. Chekhonin, eds. 2007. *Karelia: Guidebook.* Petrozavodsk: Scandinavia.

Kurets, T. 2000. *Russkie zagovory Karelii.* Petrozavodsk: PetrGU.

Kurs, O. 2001. "The Vepsians: An Administratively Divided Nationality." *Nationalities Papers* 29 (1), 69–83.

Lavonen, N. 1988. "Zagovory v krugu religiozno-magicheskikh predstavleniy karel (po materialam ekspeditsiy 1975–1984 gg.)." In *Obryady i verovaniya narodov Karelii*, edited by N. Lavonen, 130–139. Petrozavodsk: Karel'skiy nauchnyy tsentr RAN.

Mühlhäusler, P. 1996. *Linguistic Ecology: Language Change and Linguistic Imperialism in the Pacific Region.* London and New York, NY: Routledge.

Peltonen, H. 2010. "Modelling International Collective Responsibility: The Case of Grave Humanitarian Crises." *Review of International Studies* 36 (2), 239–255.

Pugh, S. M. 1999. *Systems in Contact, System in Motion: The Assimilation of Russian Verbs in the Baltic Finnic Languages of Russia.* Uppsala: Studia Uralica Upsalieansia.

Puura, U. 2012. "Veps Language: An Overview of a Language in Context." *Working Papers in European Language Diversity* 16, 1–10.

Puura, U. and O. Tanczos. 2016. "Division of Responsibility in Karelian and Veps Language Revitalization Discourse." In *Linguistic Genocide or Superdiversity? New and Old Language Diversities*, edited by R. Toivanen and J. Saarikvi, 299–325. Bristol and Buffalo, NY: Multilingual Matters.

Pyyhtinen, O. and S. Tamminen. 2011. "We Have Never Been Only Human: Foucault and Latour on the Question of the Anthropos." *Anthropological Theory* 11 (2), 135–152.

Sayer, A. 1984. *Method in Social Science. A Realist Approach.* London, Melbourne, Sydney, Auckland, Johannesburg: Hutchinson.

Schieffelin, B., K. Woolard, and P. Kroskrity, eds. 1998. *Language Ideologies: Practice and Theory.* Oxford: Oxford University Press.

Searle, J. 1969. *Speech Acts.* Cambridge: Cambridge University Press.

Setälä, E. and J. Kala. 1951. *Näytteitä Äänis- ja keskivepsän murteista.* Helsinki: Suomalais-Ugrilainen Seura.

Strogal'shchikova, Z. 2004. "Problemy stanovleniya natsional'nogo obrazovaniya karelov i vepsov na sovremennon etape." In *Sovremennoe sostoyaniye i perspektivy razvitiya karel'skogo, vepsskogo i finskogo yazykov v Respublike Kareliya*, edited by T. Kleerova and S. Pasyukova, 41–47. Petrozavodsk: Periodika.

Strogal'shchikova, Z. 2014. *Vepsy: Ocherki Istorii i Kul'tury.* Sankt-Peterburg: Inkeri.

Vakhtin, N. 2001. *Yazyki narodov severa v XX veke. Ocherki yazykovogo sdviga.* Sankt-Peterburg: Evropeyskiy Universitet v Sankt-Peterburge.

Varfolomeeva, A. 2016. "The Soul of Stone: Mineral Symbolism in Vepsian Villages of Karelia." *Laboratorium: Russian Review of Social Research* 2. Open access.. http://www.soclabo.org/index.php/laboratorium/article/view/618/1661 [accessed 24 March 2017].

Vinokurova, I. 1988. "Ritual pervogo vygona skota na pastbishcha u vepsov." *Obryady i verovaniya narodov Karelii: Sbornik statey*, 4–26.

Vinokurova, I. 2006. *Zhivotnye v traditsionnom milovozzrenii Vepsov.* Petrozavodsk: Izdatel'stvo PetrGU.

Yegorov, S. 2006. "Vepsko-Russkoe mezhetnicheskoe vzaimodeystviye." In *Sovremennaya nauka o Vepsakh: dostizheniya i perspektivy (pamyati N. I. Bogdanova)*, edited by I. Vinokurova, I. Grishina, N. Zaytseva, I. Mullonen, and V. Orfinskiy, 237–250. Petrozavodsk: Karel'skiy nauchnyy tsentr RAN.

Zaitseva, N. 2016. *Ocherki vepsskoy dialektologii (lingvogeograficheskiy aspekt)*. Petrozavodsk: Karel'skiy nauchnyy tsentr.

Archival sources

Russian Academy of Sciences (RAN), Petrozavodsk.

Zhurnal 19, kasseta 2662b, nomer 25. 1981. Pondal. Researcher: Onegina. Interviewee: Aleksandra Leont'yeva Kalinina.

12 Predators and reindeer on the same pastures?

Helena Ruotsala

Background[1]

> Oh, things have gotten so crazy. Before, we were in the reindeer forest and worked there, but now everything comes from Brussels. Today's reindeer herding is really desk-top reindeer herding, so it is.
>
> (Field notes from a discussion with a reindeer herder, April 2017)

Animals, both wild and domesticated ones, can be seen as active partners in human society, which is usually regarded as being composed solely of human beings, as, for example, Mary Weismantel and Susan Pearson have written (2010, 17). Relationships, experiences and views regarding predators and people in Finland are interesting, sometimes even conflicting. The history of the relationship between human beings and predators is long and its traces are visible. Predators can also give us a new sense of community by providing us with an interesting topic to discuss from several points of view. These multiple experiences, attitudes and views can be analysed more deeply when we look, for example, at where people live, their profession or history. Several studies have been conducted on people's attitudes towards predators and how population management and conversations about predators should be handled (e.g. Nykänen and Valkeapää 2016; Pohja-Mykrä and Kurki 2013; Rannikko 2012). The Finnish Forest and Park Service (*Metsähallitus* in Finnish), for instance, hosts an internet page called *Suurpedot* [Large Predators], where it tells people – from its own perspective – about large carnivores (Metsähallitus 2019). Some studies and surveys show that many people have both positive and negative opinions about predators. An increasingly popular type of tourism has to do with admiring and taking pictures of predators, mainly bear carcasses. There are areas of the world where predators comprise an important part of the tourist industry's resources. In Kuusamo and Kainuu, for example, watching and taking pictures of predators in their natural habitat are important parts of so-called wildlife tourism. Predators, especially bears, are tempting people to visit these areas. It is an interesting, even exciting, experience to know that there are large carnivores elsewhere; the opportunity to see traces of them or even catch a glimpse of one serve as a lure or inducement, with wildlife tourism offering people a chance to experience genuine nature. One important aspect of the positive attitudes towards large carnivores is the emphasis on biodiversity and the notion that predators are an important and integral part of Finnish nature.

In addition to positive attitudes, different negative voices can also be heard, depending very much on the background and residence of people. According to *Metsähallitus*'s web page on large carnivores, for example, "most of the negative comments stemmed from people feeling that the local communities were completely excluded from decision-making and that decisions were made solely by national entities or, at worst, by the European Union (EU) outside of Finnish borders" (Metsähallitus 2019). I have often heard these kinds of complaints during my fieldwork on reindeer herding in transition. It is one way that local people can show their power to politicians and react against the official system or majority population in some areas (Ruotsala 2002, 340).

Regarding predators, I mean wolves, bears, wolverines and lynxes. Humans have lived close to and taken predators into account both in the countryside and towns for a long time. The wolf population in Finland, for instance, had been quite large until the end of 1800s but after that, almost the entire wolf population was hunted to extinction. Hunters received a monetary bounty for every wolf they slaughtered. During the last few decades, the numbers of wolves and other predators have been increasing. People fear predators, and the latter's presence causes many to feel threatened and insecure, even in the suburbs and not only the countryside. The increasing number of predators does not impact all people positively, and strong conflicting voices have been raised, especially in the eastern parts of the country. These negative voices are understandable if we think about how rapidly the wolf population has been growing, as but one example. One very interesting aspect of the discussion regarding wolves in Finland is the notion that wolves and other predators have a nationality, as Heta Lähdesmäki (2015, 185–186) has pointed out, with their nationality being part of people's mental image of these animals.

What is reindeer herding?

A brief introduction to reindeer herding is in order to understand the questions better concerning reindeer and large carnivores inhabiting the same pastures. Reindeer have always been, in many different ways, extremely important animals in Finnish Lapland and other northern areas, for example, in Sweden, Norway and Russia. Reindeer had made life possible in the sparsely populated areas and harsh nature of Lapland for many centuries. Reindeer have provided people with food, material for clothes and dwellings, such as the kota (a tepee-like structure), and material for making objects and handicrafts. People have used reindeer for travel, used them as a decoy animal and to carry cargo. Reindeer are also a part of mythology, religion and poetry; they are the motif of fairy tales, poems and *yoiks*, a traditional form of song in Saami music performed by the Saami people of Sapmi in Northern Europe (Ryd 2007, passim.) It can be even said that life would have been very difficult, even impossible, without reindeer.

The historical perspective – where and how reindeer are used and what kinds of societies practie reindeer herding – also has an impact on the concepts of reindeer as an animal and the kinds of reindeer herding we are discussing or studying. When I wrote my dissertation on reindeer herding in transition almost 15 years ago, I discussed the differences between reindeer herding and reindeer management. One of the main differences had to do with the market economy and its premises. I did not focus only on the relationships between human beings and reindeer, although that is now my focus here. I discussed reindeer herding (*poronhoito* in Finnish), and not *porotalous*, reindeer management, where the focus is on economic issues and management as a modern profession, rather than on traditional or cultural issues. I wanted to highlight the everyday life perspective: reindeer owners' viewpoints on reindeer herding (Ruotsala 2001, 229–231; 2002, 65). Today, animal agency is widely under discussion. Bruno Latour has demanded attention be drawn to the continuous processes of social formation. Such processes are open and consist of diffuse agencies transcending the traditional limits of human actors (Ingold 2013, 12–14; Kurth 2011, 87, 93–94; Latour 2005, 76)

There are several ways to look at the history of reindeer herding while also considering the issue of predators. I will focus here briefly on Tim Ingold's (1980, 1–6) model, because the relationship between reindeer and human beings plays such an important role. Both extensive and intensive reindeer herding have also been important in the history of reindeer herding. Extensive reindeer herding is based on large pastures and free roaming reindeer. Intensive reindeer herding is based on smaller pastures but very intensive control of the reindeer, where the reindeer are used to people taking care of them, in opposition to extensive reindeer herding, where reindeer are partly without care. Regarding extensive reindeer herding, the relationship is more distant, and it is done to produce meat. The relationship can be either close or distant. Tim Ingold's old model is based on production and the relationship between animals and human beings. It tells how reindeer have been used and exploited.

In his book *Hunters, Pastoralists and Ranchers. Reindeer Economies and Their Transformations*, Tim Ingold (1980) discusses different modes of production in reindeer herding history. The first phase is hunting and, of course, taming the reindeer. The intensity of sharing in hunting societies is a key concept in this model. Sharing includes the possession and distribution of the animals hunted. Once tamed, the treating of domesticated animals as property raises new questions about the role of sharing. During this first phase, everyone had equal access to the animals, but in the next phases, the access to animals became divided and animals became private property. Though the reindeer may roam freely on other people's land or state land, with the state being the largest landowner in Finland, someone still owns them (Ingold 1980, 1–7, 78).

According to Ingold, hunting is defined by the conjunction of predatory human being–animal relationships, with subsistence production based on the principles of common access to the means of production and the sharing of produce. The second phase, pastoralism, is defined by the conjunction of protective human being–animal relationships, with the principle of divided access to animal means of production. Accumulation here involves the appropriation of the natural increase, while the production of raw materials, which entails removing animals from reproduction, is limited to the satisfaction of immediate domestic needs. Reindeer lived close to people, they were tamed, used for transport, and to produce milk and food. People moved according to the needs of the reindeer and adjusted themselves to their habits (Ingold 1980, 1–6).

Again, according to Ingold, ranching, the third phase, is defined by the predatory exploitation of animals, which, nevertheless, constitute objects of property for sale in a money market. Production for exchange, far from placing a drain on reserves of wealth, is, in this case, integral to the circuit by which it is accumulated. Capital plays a decisive role here. In ranching, the reindeer are almost wild and people no longer live with them or follow them, as in pastoralism. People do not even name them, as before (Ingold 1980, 1–6, 78).

The most common type of reindeer herding today is similar to ranching, but my point of departure here is that reindeer herding is still a way of life in the modern world. It is an occupation only for a small minority of people, but its meaning is much greater regarding cultural heritage and identity. In that sense, the question of predators and reindeer inhabiting the same pastures is even more complicated. Such encounters are very challenging and full of tension today, invoking different opinions and conflicts of interest among people. In this chapter, my aim is to shed light on the viewpoints of reindeer herders.

As stated previously, reindeer herding has been of great significance to the people living in the harsh conditions of the northern regions. It has played a decisive role in guaranteeing the viability of the sparsely populated areas precisely because reindeer can be used in so many different ways. Although it is a traditional source of livelihood in the northern parts of Finland and Scandinavia, it is a relatively recent practice. Small-scale herding, where families have a small number of tamed reindeer for use in transportation and as a decoy animal rather than for food, is hundreds of years old, however. The advent of large herds, 'reindeer capitalism', which needed different pastures at different times of the year, occurred in western parts of Finnish Lapland over the Köli mountains during the seventeenth and eighteenth centuries. For historical reasons, reindeer herding in Finland and Russia is not a privilege of just the Saami, as it has been in Sweden and in Norway. Finnish colonialists who moved to Lapland from southern parts of Finland adopted small-scale reindeer herding from the local forest Saami. In fact, most of the reindeer owners and herders in Finland are ethnically Finnish, but the significance of the practice is more pronounced in the northern parts of Lapland, where most of the herders are Saami. Reindeer herding is also regarded as one of the most important ethnic symbols of the Saami, although only a minority of them have ever been herders. One group of Saami are even called reindeer Saami. Their home areas in Finland are Enontekiö and parts of Sodankylä, Utsjoki and Inari, where other groups of Saami also live (Ruotsala 2002, 79–85; 2011, 159–160).

Reindeer herding in transition

Reindeer herding is a natural source of livelihood, and herders have had to adapt to both external and internal transformations in recent decades. Deep structural changes took place after the 1960s with the advent of the market economy. Today, it is a highly mechanised profession, which makes full use of technical equipment and rational methods of animal husbandry. The increased costs brought by technology and an adaptation to a modern way of life have changed the nature of reindeer herding. Modernisation and the increasing costs of everyday life mean that families need to have larger herds to cover all their costs. Large-scale environmental changes and other interest and user groups have put increasing pressure on the best use for diminishing pastures. Increased interests in preserving "untouched nature" has also caused reindeer and reindeer herders to come into conflict with, for example, nature protectors, scientists and politicians. Environmental changes are clearly visible in nature, and their causes are Janus-faced, making them difficult to discuss. Nature is not at all untouched in Lapland, although outsiders have difficulty seeing traces of life and people in the Lappish countryside. They emphasise that there is enough space for both reindeer and other animals in the "wildernesses" of Lapland, although the reindeer herders and locals do not share this opinion. For the latter, there is no wilderness and they do not recognise the view that culture and nature are separate. Reindeer and predators are a good example of such conflicting viewpoints (see, e.g. Haraway 2008, 3–8; Köstlin 2001, 3–6; Ruotsala 2011, 163–164).

In addition to problems with large carnivores, overgrazing or changed habits in the use of pastures also represent large changes for reindeer herding. According to some natural scientists, reindeer are no longer an essential or desired part of the fell landscape in Lapland because they disturb nature with their behaviour; they roam after food and eat the vegetation. Many scientists see modern reindeer herding as a threat to nature, especially in nature reserves, which they want to preserve as scientific laboratories where they can study rare flowers and vegetation (e.g. Nykänen and Valkeapää 2016). Reindeer herders, for their part, are wondering why certain small plants are much more important than their reindeer, which are the basis of their livelihood and have long been using the areas for grazing. In addition to the changes in reindeer herding, large predators are now part of the reindeer herding areas and politics – and, therefore, conflicts. The topic of and reasons for overgrazing problems and large predators are many-sided. Therefore, herders and their families are living in a very stressful situation today. They feel they are powerless, as one herder said: "I want to quit and let my boy continue this work. It seems that he is willing to work with reindeer, with all the positive and negative sides, to struggle against other factors and wolverines and bears. I am soon old, I don't care" (TYKL-tutkimusarkisto 1.6.27/3a, 3b). This is a quote from one of my interviews in 1999.

Predators in the reindeer forest

Metsähallitus's web page on large carnivores states that

> a majority of reindeer owners approve of having large carnivores in the reindeer herding area, provided that the damages they cause are compensated for and their populations are managed. The views on large carnivores are the most positive in the northern reindeer herding area and among the younger age groups.
>
> (Metsähallitus 2019)

My aim in this article is to study this issue based on my fieldwork, mainly interviews conducted with members of one reindeer herders' association, Kyrö, which is located in the western part of Lapland.

Lapland is often regarded as a sparsely populated area in the northern periphery, which has enough space for such predators as wolverine, wolf and bear. This stereotype of a wilderness does not fit with the viewpoint of the people who live there and herd reindeer. Reindeer herding is a very important source of livelihood, both economically and culturally. It is also a way of life for the people dependent on reindeer herding, one which is the source of much attention from other groups, such as wildlife researchers, politicians, decision makers, tourists and nature protectors. For all of them, it is important to disseminate knowledge about different animals and their living conditions. Various groups also have different needs and values, which come into conflict regarding questions about predators and reindeer (Ruotsala 2011, 164). They have never been "companion animals".[2] The discussions and conflicts about reindeer and predators are today very political and also very heated, but they are important. My main research questions here are as follows: Is there enough space for predators and reindeer in the same pastures? Who has the power to speak and decide on these issues? I will pose some answers here to these questions, which have many sides.

The question concerning predators and reindeer herding is today very vital and political in Lapland, as I heard last spring when I was attending "*Poropäivät*", a day celebrating reindeer herding in a village where I had earlier done my fieldwork. I met a former informant there, a reindeer herder, who wanted to discuss with me the current changes and issues pertaining to reindeer herding. He said that somebody should study the relationship between reindeer and predators because the situation has changed so much during the last few years. According to him, it is no longer profitable to be a reindeer herder. He then said the same thing he had told me at the end of 1990s: that the cultural values of reindeer herding are so important that he will continue with it (Field notes April 2018; TYKL-tutkimusarkisto 1.6.27/5a, 5b.).

My doctoral dissertation on reindeer herding was largely based on fieldwork conducted over the course of several years. The main point of it was to study how pastures had changed ecologically, culturally and socially from the 1930s to the mid-1990s, when Finland became a member of the EU. My aim was also to make reindeer herders' voices heard. One theme was to assess how reindeer herders and their family members felt about predators and why. Since publishing my study in 2002, I have continued to follow the changes in reindeer herding with interest, because I come from a reindeer herding family and our summer house is on the border between two different reindeer herding associations. Both the positive and negative sides of modern reindeer herding are visible there. As part of the project 'Animal Agency in Human Societies: Finnish Perspectives 1890–2040', financed by the Academy of Finland, my intention is to study the relationship between reindeer, reindeer-herding dogs and predators in reindeer herding.

One significant change between my study in 2002 and the situation today is the question concerning predators and reindeer herding. When I did my original research, the debate on predators in reindeer herding areas was not so heated – and not found throughout Finland, as it is now. One aspect of the recent debate has to do with the changes in the protection of predators and their numbers. The wolf population, for instance, was severely threatened until the 1970s when they became a protected species. It was also protected in other parts of Finland outside the reindeer herding area. One reason for the differences between my old study on reindeer herding and the current situation is the fact that my study ended when Finland became a member of EU. This membership has brought with it some changes in the politics of protecting nature and predators. Finland is no longer responsible on its own for its predator politics; it also has to consider the opinions of other EU member states. This caused a lot of talk and herders were very worried about it (Ruotsala 2002, 404–405).

One issue concerning predators is the different areas in which reindeer herding takes place, which differ from other areas in Finland. The reindeer herding area in Finland accounts for 36 per cent of the nation's total area, but most of the reindeer today only live in the northernmost areas. There are approximately 150,000–200,000 reindeer, but the numbers are controlled by the people involved and by a particular reindeer herding area, the *paliskunta*. Different predators live in different areas; in my home area, for example, there are hardly any wolves, but the eastern part of the country, near the border with Russia, is 'suffering' from an excess number of wolves. Bears are the most common predator in the reindeer herding area and become a real threat during spring and summertime and especially for the newborn calves, which are a prime target. The most common threat – and also dangerous from a reindeer herders' point of view – is the wolverine. It also causes most of the damage if we look at the Finnish reindeer herding area as a whole. It accounts for more damage than the other three predators combined (Pyykkö 2018, 4–5). The wolverine is a very strong and tough animal and very fast. It kills especially pregnant reindeer in the springtime – when the snow is hard and it can run very fast. However, the wolverine is a strictly protected animal. "We don't like the wolverine so much, and now in the springtime we are afraid of it. It kills for fun, not only for hunger" (Field notes, April 2017).

Reindeer differ from wild animals because they are always somebody's property, even if they live freely in nature. They roam free for most of the year and herders only round them up several times a year.

Predators have always been a natural part of the reindeer forest, thus, the relationship between animals and reindeer herders has never been without conflicts and losses. The role of humans has been to herd reindeer in order to protect them from predators. The reindeer dog has been an important helper in such a task. Work in the reindeer forest has changed a great deal during the last 10 years for other reasons as well, and the increasing amount of damage caused by predators is one of them. The work itself and its cultural and symbolic value are important, but predators harm and disturb the work – in some areas too much. Some herders even say that the work in some areas is mainly to prevent or else keep track of the damage caused by predators, and this is regarded as the most important threat to their way of life (Nykänen and Valkeapää 2016, 44–45). The predator population differs in various reindeer herding areas: wolves are the biggest threat in some places, while in other areas, wolverines or bears threaten the reindeer the most. The amount of damage caused to reindeer observed correlates strongly with the number of predators in the same areas. Thus, in areas where, for example, the wolf or wolverine populations are higher, the number of reindeer they kill is also higher. Reindeer herders also know that in addition to the number of large carnivores, weather and other natural conditions, such as the quality and quantity of snow and the quality of pastures, are also important factors in the damage caused to reindeer (TYKL-tutkimusarkisto 1.6.27/5b, 7, 8, 14b, 23; Ruotsala 2002, 338–339).

As already stated earlier, herding reindeer is different in various areas. In my research area, the Kyrö Reindeer Herder's Association, for instance, reindeer are herded from September to December, in April, and then again from the middle of June to the middle of July. The reindeer roam freely during other times of the year. Reindeer are also kept behind fences in some places during the winter and spring months, so they can be given extra food because the pastures are in bad condition in those regions during those times of year. Snowmobiles, four-wheeled vehicles, helicopters and even drones (small helicopter cameras) are used in herding. Herders keep track of the reindeer and their movements and ensure that the herd stays in a certain area and, of course, observe where large predators are moving; a need which has increased in the last few years. Through herding it is possible to prevent and ascertain the losses caused by predators. Some reindeer owners even keep their reindeer behind fences for most of the year, except the summer months, to prevent the damage caused by large carnivores.

According to reindeer herders, wolverines are now the largest threat in the area where I did my interviews. Bears mostly disturb the reindeer during the calving period and at the beginning of summer. Herders say that it is difficult to find the calves killed by bears because they eat them and hide them under trees and bushes. During the summer months, a small calf can easily disappear. "Well, it is so if the bear enters a herd where there are many small calves, it will take the calves", as one old herder told me (TYKL-tutkimusarkisto 1.6.27/2).

Figure 12.1, provided by the Reindeer Herders' Association, shows the number of reindeer that have been killed by large carnivores in Finland between 1991 and 2016. Reindeer herders are compensated for the reindeer killed by large carnivores in accordance with the Game Animal Damages Act. However, not all reindeer that have been killed by large carnivores can be found; according to this picture, only 1 in 5 are found. The picture shows how the number of reindeer killed by predators has increased from the beginning of 1990s. The number of reindeer killed (and found) in 1993 was 1000, whereas in 2016 it was 4885. What is also interesting is that the number of reindeer killed by large carnivores varies in different parts of the total area managed by this large reindeer herders' association.

Why are predators so great a threat in reindeer herding? In some areas, especially in the eastern and southern parts of Lapland, the threat is so great that local people think reindeer herding is no longer economically feasible,

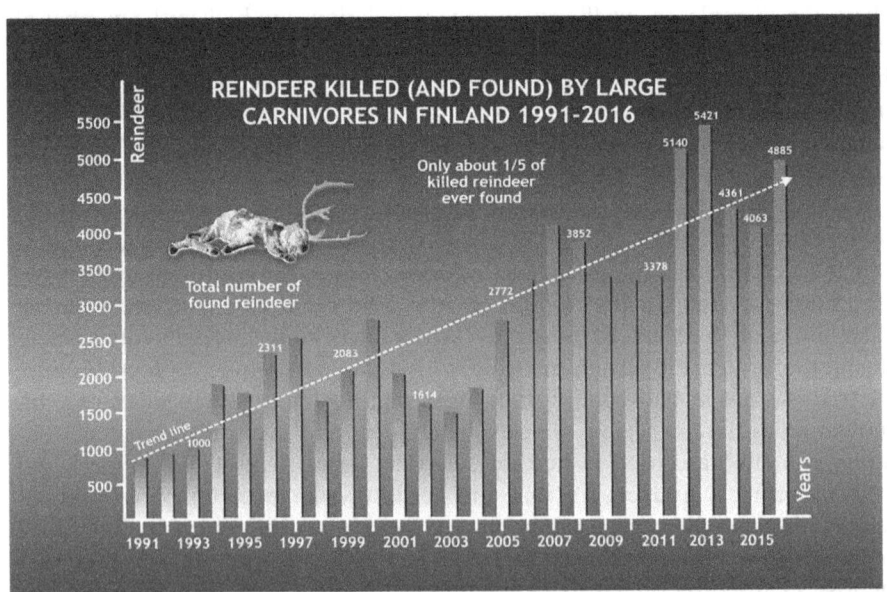

Figure 12.1 Reindeer killed (and found) by large carnivores in Finland, 1991–2016
Source: Reindeer Herders' Association, 2017

which also has an effect on cultural values. For these people, large predators are the greatest threat to reindeer herding today (Pakkanen and Valkonen 2012, 44–45).

Figure 12.2 shows clearly how the amount of damage caused to reindeer by predator animals has been increasing of late. The state budget sets limits for the total compensation allotted to reindeer owners for their losses. The compensation is only paid for reindeer that have been found and where it is clear that they were killed by large carnivores. Consequently, the amount of compensation for a found reindeer is 1.5 times the animal's value. The carcasses are recognised by their earmark or collar, which ascertain whose reindeer has died. The owners of reindeer that have died, for example, from starvation, or due to snowy conditions or traffic, do not receive such compensation. Damage caused by vehicles are paid for by different traffic insurances. On the one hand, reindeer owners say that not all animals killed by large carnivores can be found, while, on the other hand, reindeer owners are criticised for receiving too much compensation – also for animals that have died for other reasons. Reindeer owners also blame the politicians – namely, the Ministry of Agriculture and Forestry – for the small amount of compensation, because their losses in the last few years have exceeded the limit many times over. The number of reindeer killed by predators has been increasing so quickly that the state budget no longer covers the costs to reindeer herders (Field notes April 2017; Pyykkö 2018, 4–5).

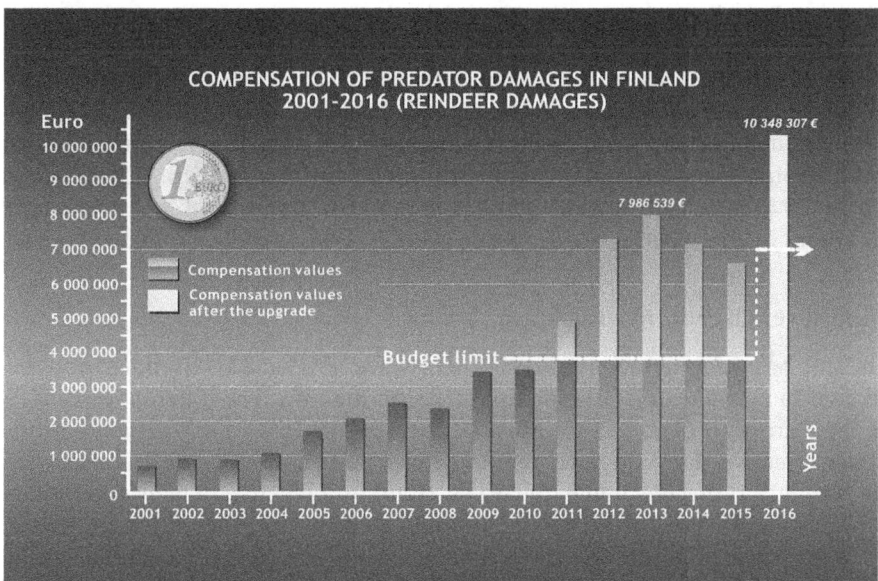

Figure 12.2 Compensation for damage caused by predators in Finland, 2001–2016
Source: Reindeer Owners' Association

In spring 2018, a local newspaper published a list of reindeer herders who had received the most compensation for reindeer killed by predators and vehicles. The article sparked a heated debate, but the reindeer herders who were on the list acquired a negative reputation. It caused them a great deal of stress and even fear, because, according to them, the compensation is related to the hazards of the job and is not extra money, as was stated in the discussion (Pyykkö 2018, 4–5).

Although the damage caused by predators is compensated for by the state, according to my informants, the system falsifies or imperils the economic meaning of reindeer herding. These statistics are skewed by inflation, because the compensation or the money received as a result of losses is not from the herding work. For the herders, tending to dead animals and informing the civil servants about them is not proper work, although they need the money. In the long run, this scenario – with compensation paid by the state – means that herders cannot breed the reindeer, which – according to herders – is the proper work done in a reindeer forest. There is no future in the work – or so some especially younger herders told me. As one reindeer herder put it: "We don't want to merely raise reindeer for the predators [to kill], but now [April] it seems to be so, when the pregnant female reindeer are eaten by wolverines" (Field notes, April 2017).

There is a widespread desire that the populations of large predators should be more evenly distributed throughout Finland, but this is not yet the case. The increasing wolf population, especially in the areas of the reindeer herders' associations of eastern Lapland, has had a deep impact on the number of reindeer. The threat of damage caused by large carnivores creates insecurity and fear. The damage caused by wolves is so significant that local people even staged demonstrations in the spring with organised funerals for the reindeer who had been found after a wolf had killed them (Karvinen 2012). However, it is interesting that, at least according to some surveys, the fear of wolves is much stronger in western and southern Finland than in northern or even eastern Finland, where the human population is much smaller and it is possible to find more open expanses of wilderness (Pohja-Mykrä and Kurki 2013, 11–13).

Finnish reindeer in the European Union

Finland has been a member of the EU since 1995, and this must now be taken into account in legislation and decision-making. The different directives of the European Commission must be incorporated into national legislation. Many directives have targeted reindeer herders. Exemplarily, large carnivores are strongly protected under the EU's Habitats Directive, which determines how various Finnish parties should view the relationship between predators and reindeer herding. It allows for certain exceptions regarding hunting, about which I will not go into detail here. Suffice to say, it is possible to make certain exceptions to prevent damage to, for example, reindeer. This has been very hard for reindeer owners, who know more about the damage caused by wolverines. The Habitats Directive does not regulate wolverine hunting, but it will most probably be regulated soon. The problem of wolves in the reindeer herding areas has not been heavily regulated thus far either (Borgström 2011, 144–150; Paliskuntain yhdistys 2017; Pohja-Mykrä and Kurki 2013, 11).

This EU Habitats Directive is not viewed only in a positive light by reindeer herders, although its purpose has been to promote and support biodiversity and sustainable development. Finland has a strict protection system in place for bears, wolves and lynxes. The local people and reindeer herders do not always share the opinions of outsiders about which species are rare and endangered and, therefore, must be protected in their area. Reindeer herders blamed Finnish negotiators when Finland became a member of the EU for not seeking better solutions that would consider their source of livelihood more. One old reindeer herder put it as follows:

> You see it that the first negotiators didn't know the meaning of reindeer herding for us, who live here. It is our source of living; we have to go to the reindeer forest every day and check whether we see the predators and what is left [of the reindeer] after them. It is different behind a desk than out here.
> (TYKL-tutkimusarkisto 1.6.27/3b, 16; Ruotsala 2002, 405)

One important issue is that a herder must own a minimum of 80 reindeer in order to obtain EU compensation for reindeer that have been killed. If the losses caused by predators are large, the herder must then keep the remaining stock alive – including the bad ones, which should not otherwise be used for breeding. The compensation for damages is necessary for the herders; today, it has become so necessary that some even regard it as an essential part of their income. For others, it is only compensation for damage. To receive compensation for any dead animal killed by predators, the herder must find them to verify how they died. It means extra work, which takes away from herding and other important tasks. Some of the herders are now using collars on the reindeer, and the collars send a signal when the reindeer stops moving. Therefore, large predators cause extra work and time and extra costs: herding reindeer in such a way as to avoid additional damages, finding animals that have already been killed and buying high-tech collars, not to mention filling in more and more forms and applications to receive compensation, as my informants told me when emphasising the changes. This is what reindeer herders dismissively call "desktop reindeer herding" (e.g. Ruotsala 2002, 404–405).

I have not discussed the killing of predators here, which is currently officially a strictly regulated system. There are licences for how many and where and when you can shoot predators – herders complain that this system moves very slowly; in the spring of 2015, for example, when a bear made its way past a calving fence, it took many days to obtain the licence to kill it and, by then, it had eaten almost all the calves (Palomaa 2018, 10; Field notes, April 2017).

The illegal killing of predators and the anger generated by such actions are issues that are – also stereotypically – connected with reindeer herding. They are not discussed in public. Reindeer herding associations are currently working to combat this anger. It is extremely difficult to discuss issues such as the illegal hunting of predators in a rational manner because they fall under criminal law. The ethical questions are so pressing that it is difficult even to try to speak about them (see Pohja-Mykrä and Kurki 2013, 15–17). Hunting predators without a licence can be seen as a reaction against a perceived position of powerlessness, where regulations and rules are decided by the EU, which – according to my interviews – does not listen so much to local people.

The coexistence of reindeer and predators

Predators and reindeer have always occupied the same pastures, but reindeer herders say that today the problem concerning predators is increasingly a political one. In addition, more and more different interest groups have arrived to take part in the discussions concerning reindeer and predators. While a certain number of predators is acceptable, how will it be possible to identify the acceptable or reasonable number of predators in a particular reindeer herding area? A large number of wolves and bears, for instance, is not acceptable in southern Finland, where there are no reindeer. We can read in the media how pupils in southern parts of Finland fear wolves so much that they go to school in cars, not on foot. They are even afraid to go to the neighbouring houses because of the threat of wolves, which can enter people's gardens (see, e.g. *Kainuun Sanomat* 2017).[3] Herders must rely on their own officers, who have negotiated with other partners in the EU regarding the number of predators and how to avoid the damages they are causing in reindeer herding areas. But reindeer herders think that the question concerning large carnivores in reindeer herding areas needs to be resolved because it constitutes the main economic problem for people working in this occupation and strongly affects the way of life in some areas. It also increases the level of stress that other interest groups, such as nature protection organisations, nature lovers, scientists and some politicians, are causing for reindeer herders. How to cope with media is also a problem for reindeer herders in such a stressful situation.

The coexistence of reindeer and large carnivores has never been without its problems, but it is totally possible that all can find room in the reindeer pastures. Reindeer and large carnivores are an integral part of Finnish nature and are important for biodiversity. Predators – but especially reindeer – are an important part of the Finnish image and used to market Finland and Finnish wilderness areas to foreigners.

Notes

1 This article belongs to the research project "Animal Agency in Human Society" funded by the Academy of Finland (274573).
2 For more on companion animals, see Haraway (2003, 100).
3 "*Ministeri Jari Leppä laittaisi laskun susien takia ajetuista kyydeistä komissiolle*", [Minister Jari Leppä would send the bill concerning school taxi caused by the wolves to the EU Commission].

Bibliography

Borgström, S. 2011. *Iso paha susi vai hyödyllinen hukka? Ekologis-juridinen näkökulma suden suojelun yhteiskunnalliseen hyväksyttävyyteen.* Publications of the University of Eastern Finland. Dissertations in Social Sciences and Business Studies, no 20. Joensuu: Eastern Finland University.

Haraway, D. 2003. *The Companion Species Manifesto: Dogs, People, and Significant Otherness.* Chicago, IL: Prickly Paradigm Press.

Haraway, D. 2008. *When Species Meet.* Minneapolis, MN: University of Minnesota Press.

Ingold, T. 1980. *Hunters, Pastoralists and Ranchers. Reindeer Economy and Their Transformations.* Cambridge: Cambridge University Press.

Ingold, T. 2013. "Anthropology Beyond Humanity. Edward Westermarch Memorial Lecture, May 2013." *Suomen Antropologi: Journal of the Finnish Anthropological Society* 28 (3), 5–23.

Kainuun Sanomat, 2017. "Ministeri Jari Leppä laittaisi laskun susien takia ajetuista koulukyydeistä komissiolle." *Kainuun Sanomat.* 28 November.

Karvinen, T. 2012. "Petojen syömät porot herättävät närää." *Maaseudun Tulevaisuus.* 29 April.

Köstlin, K. 2001. "Kultur als Natur – des Menschen." In *Natur – Kultur. Volkskundliche Perspektiven auf Mensch und Umwelt.* 32. Kongress der Deutschen Gesellschaft für Volkskunde in Halle vom 27.9. bis 1.10.1999, edited by R. W. Brednich et al., 1–10. New York, NY, München and Berlin: Waxmann Münster.

Kurth, M. 2011. "Von mächtigen Repräsentationen und ungehörten Artikulationen. Die Sprache der Mecsh-Tier-Verhältnisse." In *Human-Animal Studies. Über die gesellschaftliche Natur von Mesch-Tier-Verthälnisse,* edited by Chimara Arbeitskreis, 85–119. Bielefelt: Transcript.

Lähdesmäki, H. 2015. "Susi yhteiskunnallisena eläimenä." In *Eläimet yhteiskunnassa,* edited by E. Aaltola and S. Keto, 185–191. Helsinki: Into.

Latour, B. 2005. *Reassembling the Social: An Introduction to Action-Network Theory.* Oxford: Oxford University Press.

Metsähallitus [Finnish Forest and Park Service], 2019. "Large Carnivores." Suurpedot. fi. http://www.largecarnivores.fi/front-page.html

Nykänen, T. and L. Valkeapää. 2016. *Kilpisjärven poliittinen luonto – matkoja Käsivarren kulttuurimaisemassa.* Helsinki: Suomalaisen Kirjallisuuden Seura.

Pakkanen, A. and J. Valkonen. 2012. "Porotalouden hyvinvointi ja tulevaisuuskuvat eteläisissä paliskunnissa." In *Poronhoitajien hyvinvoinnin uhat ja avun tarpeet,* edited by A. Pohjola and J. Valkonen, 13–65. Rovaniemi: Lapland University Press.

Paliskuntain yhdistys: Reindeer Herder's Association2017. "Homepage." https://paliskunnat.fi/reindeer-herders-association/.

Palomaa, M. 2018. "Kittilään yksi ahmalupa." *Kittilälehti.* 17 January, 1, 10.

Pohja-Mykrä, M. and S. Kurki. 2013. *Suurpetopolitiikka kriisissä – salakaadot ja yhteisön tuki.* Raportteja 98. Helsinki: Ruralia-instituutti, Helsingin yliopisto.

Pyykkö, R. 2018. "Petovahingot paisuivat ennätyksen: seitsemän poronomistajaa sai korvauksena vähintään 100 000 euroa, suurin summa ylitti 218 000." *Lapin Kansa.* 5 April, 4–5.

Rannikko, P. 2012. "Susien suojelun tragedia: autoetnografinen tutkimus salametsästyksen paikallisesta hyväksyttävyydestä." *Alue ja ympäristö* 42 (2), 70–80.

Ruotsala, H. 2001. "Rentiere, die 'Rinder' des Nordens im Naturpark." In *Natur – Kultur. Volkskundliche Perspektiven auf Mensch und Umwelt*. 32. Kongress der Deutschen Gesellschaft für Volkskunde in Halle vom 27.9. bis 1.10.1999, edited by R. W. Brednich et al., 229–238. New York, NY, München and Berlin: Waxmann Münster.

Ruotsala, H. 2002. *Muuttuvat palkiset. Elo, työ ja ympäristö Kittilän Kyrön paliskunnassa ja Kuolan Luujärven poronhoitokollektiiveissa 1930–1995* [English abstract: Reindeer herding in transition. Reindeer management in Finnish Lapland and on the Kola Peninsula in Russia 1930–1995]. Kansatieteellinen Arkisto 49. Helsinki: Suomen Muinaismuistoyhdistys.

Ruotsala, H. 2011. "Ancestors' Wisdom or Desktop Reindeer Management? The Role of Traditional Ecological Knowledge in Contemporary Reindeer Herding." In *Thinking Through the Environment: Green Approaches to Global History*, edited by T. Myllyntaus, 159–178. Cambridge: White Horse.

Ryd, Y. 2007. *Ren och varg. Samer berättar*. Stockholm: Natur & kultur.

Weismantel, M. and S. Pearson. 2010. "Does 'The Animal' Exist? Toward a Theory of Social Life with Animals." In *Beastly Natures: Animals, Humans, and the Study of History*, edited by D. Bratz, 17–37. Charlottesville, VA: University of Virginia Press.

Research material

Archival sources: The Archives of History, Culture and Arts Studies, TYKL collection of European Ethnology, University of Turku. TYKL-tutkimusarkisto 1.6.27/1–29; Interviews and field notes made by Helena Ruotsala.

Field notes in April 2017 (not yet archived), made by Helena Ruotsala.

Index